Series

Putting **Essential Understanding** of
Statistics
into Practice

in Grades
9–12

Terry Crites
Northern Arizona University
Flagstaff, Arizona

Roy St. Laurent
Northern Arizona University
Flagstaff, Arizona

Terry Crites
Volume Editor
Northern Arizona University
Flagstaff, Arizona

Barbara J. Dougherty
Series Editor
University of Missouri
Columbia, Missouri

NATIONAL COUNCIL OF
TEACHERS OF MATHEMATICS

more**4u**
www.nctm.org/more4u
Access code: STA14547

Library of Congress Cataloging-in-Publication Data

Crites, Terry.
  Putting essential understanding of statistics into practice in grades 9–12 / Terry Crites,
Northern Arizona University, Flagstaff, Arizona, Roy St. Laurent, Northern Arizona
University, Flagstaff, Arizona.
      pages   cm.   − (Putting essential understanding into practice series)
  ISBN 978-0-87353-737-7
1. Mathematical statistics–Study and teaching (Secondary) 2. Statistics–Study and
teaching (Secondary) I. St. Laurent, Roy Thomas
II. Title.
  QA276.18.C75 2015
  519.5071–dc23

                           2014044716

The National Council of Teachers of Mathematics is the public voice of mathematics education, providing vision, leadership, and professional development to support teachers in ensuring equitable mathematics learning of the highest quality for all students.

Printed in the United States of America

Contents

Chapter 3
Data Collection and Hypothesis Testing 59

Chapter 4
Estimators 91

Chapter 5
Looking Back and Looking Ahead with Statistics 107

Appendix 1

Appendix 2

Appendix 3

References

Accompanying Materials at More4U

Appendix 1

The Big Ideas and Essential Understandings for Statistics

Appendix 2

Resources for Teachers

Foreword

Teaching mathematics in prekindergarten–grade 12 requires knowledge of mathematical content and developmentally appropriate pedagogical knowledge to provide students with experiences that help them learn mathematics with understanding, while they reason about and make sense of the ideas that they encounter.

In 2010 the National Council of Teachers of Mathematics (NCTM) published the first book in the Essential Understanding Series, focusing on topics that are critical to the mathematical development of students but often difficult to teach. Written to deepen teachers' understanding of key mathematical ideas and to examine those ideas in multiple ways, the Essential Understanding Series was designed to fill in gaps and extend teachers' understanding by providing a detailed survey of the big ideas and the essential understandings related to particular topics in mathematics.

The Putting Essential Understanding into Practice Series builds on the Essential Understanding Series by extending the focus to classroom practice. These books center on the pedagogical knowledge that teachers must have to help students master the big ideas and essential understandings at developmentally appropriate levels.

To help students develop deeper understanding, teachers must have skills that go beyond knowledge of content. The authors demonstrate that for teachers—

- understanding student misconceptions is critical and helps in planning instruction;

- knowing the mathematical content is not enough—understanding student learning and knowing different ways of teaching a topic are indispensable;

- constructing a task is important because the way in which a task is constructed can aid in mediating or negotiating student misconceptions by providing opportunities to identify those misconceptions and determine how to address them.

Through detailed analysis of samples of student work, emphasis on the need to understand student thinking, suggestions for follow-up tasks with the potential to move students forward, and ideas for assessment, the Putting Essential Understanding into Practice Series demonstrates best practice for developing students' understanding of mathematics.

The ideas and understandings that the Putting Essential Understanding into Practice Series highlights for student mastery are also embodied in the Common Core State

Standards for Mathematics, and connections with these new standards are noted throughout each book.

On behalf of the Board of Directors of NCTM, I offer sincere thanks to everyone who has helped to make this new series possible. Special thanks go to Barbara J. Dougherty for her leadership as series editor and to all the authors for their work on the Putting Essential Understanding into Practice Series. I join the project team in welcoming you to this special series and extending best wishes for your ongoing enjoyment—and for the continuing benefits for you and your students—as you explore Putting Essential Understanding into Practice!

Linda M. Gojak
President, 2012–2014
National Council of Teachers of Mathematics

Preface

The Putting Essential Understanding into Practice Series explores the teaching of mathematics topics in K–grade 12 that are difficult to learn and to teach. Each volume in this series focuses on specific content from one volume in NCTM's Essential Understanding Series and links it to ways in which those ideas can be taught successfully in the classroom.

Thus, this series builds on the earlier series, which aimed to present the mathematics that teachers need to know and understand well to teach challenging topics successfully to their students. Each of the earlier books identified and examined the big ideas related to the topic, as well as the "essential understandings"—the associated smaller, and often more concrete, concepts that compose each big idea.

Taking the next step, the Putting Essential Understanding into Practice Series shifts the focus to the specialized pedagogical knowledge that teachers need to teach those big ideas and essential understandings effectively in their classrooms. The Introduction to each volume details the nature of the complex, substantive knowledge that is the focus of these books—*pedagogical content knowledge*. For the topics explored in these books, this knowledge is both student centered and focused on teaching mathematics through problem solving.

Each book then puts big ideas and essential understandings related to the topic under a high-powered teaching lens, showing in fine detail how they might be presented, developed, and assessed in the classroom. Specific tasks, classroom vignettes, and samples of student work serve to illustrate possible ways of introducing students to the ideas in ways that will enable students not only to make sense of them now but also to build on them in the future. Items for readers' reflection appear throughout and offer teachers additional opportunities for professional development.

The final chapter of each book looks at earlier and later instruction on the topic. A look back highlights effective teaching that lays the earlier foundations that students are expected to bring to the current grades, where they solidify and build on previous learning. A look ahead reveals how high-quality teaching can expand students' understanding when they move to more advanced levels.

Each volume in the Putting Essential Understanding into Practice Series also includes appendixes that list the big ideas and essential understandings related to the topic, detail resources for teachers, and present the tasks discussed in the book. These materials, which are available to readers both in the book and online at www.nctm.org/more4u, are intended to extend and enrich readers' experiences and possibilities for using the book. Readers can gain online access to these materials by going to the More4U website and entering the code that appears on the book's title page. They can then print out these materials for personal or classroom use.

Because the topics chosen for both the earlier Essential Understanding Series and this successor series represent areas of mathematics that are widely regarded as challenging to teach and to learn, we believe that these books fill a tangible need for teachers. We hope that as you move through the tasks and consider the associated classroom implementations, you will find a variety of ideas to support your teaching and your students' learning.

Introduction

Shulman (1986, 1987) identified seven knowledge bases that influence teaching:

1. Content knowledge

2. General pedagogical knowledge

3. Curriculum knowledge

4. Knowledge of learners and their characteristics

5. Knowledge of educational contexts

6. Knowledge of educational ends, purposes, and values

7. Pedagogical content knowledge

The specialized content knowledge that you use to transform your understanding of mathematics content into ways of teaching is what Shulman identified as item 7 on this list—*pedagogical content knowledge* (Shulman 1986). This is the knowledge that is the focus of this book—and all the volumes in the Putting Essential Understanding into Practice Series.

Pedagogical Content Knowledge

In mathematics teaching, pedagogical content knowledge includes at least four indispensable components:

1. Knowledge of curriculum for mathematics

2. Knowledge of assessments for mathematics

3. Knowledge of instructional strategies for mathematics

4. Knowledge of student understanding of mathematics (Magnusson, Krajcik, and Borko 1999)

These four components are linked in significant ways to the content that you teach.

Even though it is important for you to consider how to structure lessons, deciding what group and class management techniques you will use, how you will allocate time, and what will be the general flow of the lesson, Shulman (1986) noted that it is even more important to consider *what* is taught and the *way* in which it is taught. Every day, you make at least five essential decisions as you determine—

1. which explanations to offer (or not);

2. which representations of the mathematics to use;

3. what types of questions to ask;

4. what depth to expect in responses from students to the questions posed; and

5. how to deal with students' misunderstandings when these become evident in their responses.

Your pedagogical content knowledge is the unique blending of your content expertise and your skill in pedagogy to create a knowledge base that allows you to make robust instructional decisions. Shulman (1986, p. 9) defined pedagogical content knowledge as "a second kind of content knowledge…, which goes beyond knowledge of the subject matter per se to the dimension of subject matter knowledge *for teaching*." He explained further:

> Pedagogical content knowledge also includes an understanding of what makes the learning of specific topics easy or difficult: the conceptions and preconceptions that students of different ages and backgrounds bring with them to the learning of those most frequently taught topics and lessons. (p. 9)

If you consider the five decision areas identified at the top of the page, you will note that each of these requires knowledge of the mathematical content and the associated pedagogy. For example, teaching statistics requires that you understand the need to work with a random sample from a population and then provide your students with contextual situations that help them recognize this need. Your knowledge of statistics can help you craft tasks and questions that provide counter-examples and ways to guide your students in seeing connections within and among multiple contexts. As you establish the content, complete with learning goals, you then need to consider how to move your students from their initial understandings to deeper ones, building rich connections along the way.

The instructional sequence that you design to meet student learning goals has to take into consideration the misconceptions and misunderstandings that you might expect to encounter (along with the strategies that you expect to use to negotiate them), your expectation of the level of difficulty of the topic for your students, the progression of experiences in which your students will engage, appropriate collections of representations for the content, and relationships between and among statistics and other topics.

Model of Teacher Knowledge

Grossman (1990) extended Shulman's ideas to create a model of teacher knowledge with four domains (see fig. 0.1):

1. Subject-matter knowledge

2. General pedagogical knowledge

3. Pedagogical content knowledge

4. Knowledge of context

Subject-matter knowledge includes mathematical facts, concepts, rules, and relationships among concepts. Your understanding of the mathematics affects the way in which you teach the content—the ideas that you emphasize, the ones that you do not, particular algorithms that you use, and so on (Hill, Rowan, and Ball 2005).

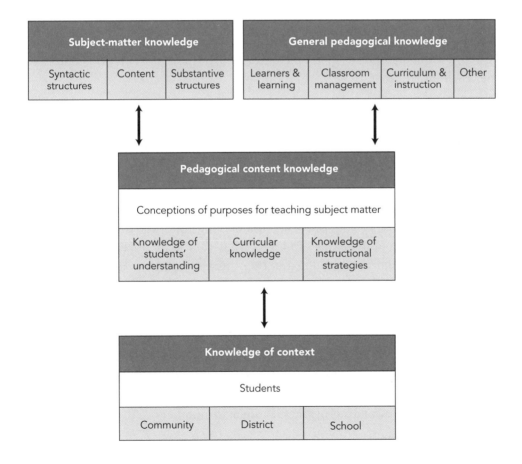

Fig. 0.1. Grossman's (1990, p. 5) model of teacher knowledge

Your pedagogical knowledge relates to the general knowledge, beliefs, and skills that you possess about instructional practices. These include specific instructional strategies that you use, the amount of wait time that you allow for students' responses to questions or tasks, classroom management techniques that you use for setting expectations and organizing students, and your grouping techniques, which might include having your students work individually or cooperatively or collaboratively, in groups or pairs. As Grossman's model indicates, your understanding and interpretation of the environment of your school, district, and community can also have an impact on the way in which you teach a topic.

Note that pedagogical content knowledge has four aspects, or components, in Grossman's (1990) model:

1. Conceptions of purposes for teaching

2. Knowledge of students' understanding

3. Knowledge of curriculum

4. Knowledge of instructional strategies

Each of these components has specific connections to the classroom. It is useful to consider each one in turn.

First, when you think about the goals that you want to establish for your instruction, you are focusing on your conceptions of the purposes for teaching. This is a broad category but an important one because the goals that you set will define learning outcomes for your students. These conceptions influence the other three components of pedagogical content knowledge. Hence, they appropriately occupy their overarching position in the model.

Second, your knowledge of your students' understanding of the mathematics content is central to good teaching. To know what your students understand, you must focus on both their conceptions and their misconceptions. As teachers, we all recognize that students develop naïve understandings that may or may not be immediately evident to us in their work or discourse. These can become deep-rooted misconceptions that are not simply errors that students make. Misconceptions may include incorrect generalizations that students have developed, such as the erroneous idea that a sample, regardless of its size, will always have the same characteristics as the population from which it was drawn. These generalizations may even be predictable notions that students exhibit as part of a developmental trajectory, such as a lack of understanding of the law of small numbers.

Part of your responsibility as a teacher is to present tasks or to ask questions that can bring misconceptions to the forefront. Once you become aware of misconceptions in students' thinking, you then have to determine the next instructional steps. The mathematical ideas presented in this volume focus on common misconceptions that students form in relation to a specific topic–statistics in grades 9–12. This book shows how the type of task selected and the sequencing of carefully developed questions can bring the misconceptions to light, as well as how particular teachers took the next instructional steps to challenge the students' misconceptions.

Third, curricular knowledge for mathematics includes multiple areas. Your teaching may be guided by a set of standards such as the Common Core State Standards for Mathematics (CCSSM; National Governors Association Center for Best Practices and Council of Chief State School Officers 2010) or other provincial, state, or local standards. You may in fact use these standards as the learning outcomes for your students. Your textbook is another source that may influence your instruction. With any textbook also comes a particular philosophical view of mathematics, mathematics teaching, and student learning. Your awareness and understanding of the curricular perspectives related to the choice of standards and the selection of a textbook can help to determine how you actually enact your curriculum. Moreover, your district or school may have a pacing guide that influences your delivery of the curriculum. In this book, we can focus only on the alignment of the topics presented with broader curricular perspectives, such as CCSSM. However, your own understanding of and expertise with your other curricular resources, coupled with the parameters defined by the expected student outcomes from standards documents, can provide the specificity that you need for your classroom.

In addition to your day-to-day instructional decisions, you make daily decisions about which tasks from curricular materials you can use without adaptation, which tasks you will need to adapt, and which tasks you will need to create on your own. Once you select or develop meaningful, high-quality tasks and use them in your mathematics lesson, you have launched what Yinger (1988) called "a three-way conversation between teacher, student, and problem" (p. 86). This process is not simple—it is complex because how students respond to the problem or task is directly linked to your next instructional move. That means that you have to plan multiple instructional paths to choose among as students respond to those tasks.

Knowledge of the curriculum goes beyond the curricular materials that you use. You also consider the mathematical knowledge that students bring with them from grade 8 and what they should learn by the end of grade 12. The way in which you teach a foundational concept or skill has an impact on the way in which students will interact with and learn later related content. For example, the types of representations that you include in your introduction of statistics are the ones that your students will use to evaluate other representations and ideas in later grades.

Fourth, knowledge of instructional strategies is essential to pedagogical content knowledge. Having a wide array of instructional strategies for teaching mathematics is central to effective teaching and learning. Instructional strategies, along with knowledge of the curriculum, may include the selection of mathematical tasks, together with the way in which those tasks will be enacted in the classroom. Instructional strategies may also include the way in which the mathematical content will be structured for students. You may have very specific ways of thinking about how you will structure your presentation of a mathematical idea—not only how you will sequence the introduction and development of the idea, but also how you will present that idea to your students. Which examples should you select, and which questions should you ask? What representations should you use? Your knowledge of instructional strategies, coupled with your knowledge of your curriculum, permits you to align the selected mathematical tasks closely with the way in which your students perform those tasks in your classroom.

The instructional approach in this volume combines a student-centered perspective with an approach to mathematics through problem solving. A student-centered approach is characterized by a shared focus on student and teacher conversations, including interactions among students. Students who learn through such an approach are active in the learning process and develop ways of evaluating their own work and one another's in concert with the teacher's evaluation.

Teaching through problem solving makes tasks or problems the core of mathematics teaching and learning. The introduction to a new topic consists of a task that students work through, drawing on their previous knowledge while connecting it with new ideas. After students have explored the introductory task (or tasks), their consideration of solution methods, the uniqueness or multiplicity of solutions, and extensions of the task create rich opportunities for discussion and the development of specific mathematical concepts and skills.

By combining the two approaches, teachers create a dynamic, interactive, and engaging classroom environment for their students. This type of environment promotes the ability of students to demonstrate CCSSM's Standards for Mathematical Practice while learning the mathematics at a deep level.

The chapters that follow will show that instructional sequences embed all the characteristics of knowledge of instructional strategies that Grossman (1990) identifies. One component that is not explicit in Grossman's model but is included in a model developed by Magnusson, Krajcik, and Borko (1999) is the knowledge of assessment. Your knowledge of assessment in mathematics plays an important role in guiding your instructional decision-making process.

There are different types of assessments, each of which can influence the evidence that you collect as well as your view of what students know (or don't know) and how they know what they do. Your interpretation of what students know is also related to your view of what constitutes "knowing" in mathematics. As you examine the tasks, classroom vignettes, and samples of student work in this volume, you will notice that teacher questioning permits formative assessment that supplies information that spans both conceptual and procedural aspects of understanding. *Formative assessment*, as this book uses the term, refers to an appraisal that occurs during an instructional segment, with the aim of adjusting instruction to meet the needs of students more effectively (Popham 2006). Formative assessment does not always require a paper-and-pencil product but may include questions that you ask or tasks that students complete during class.

The information that you gain from student responses can provide you with feedback that guides the instructional flow, while giving you a sense of how deeply (or superficially) your students understand a particular idea—or whether they hold a misconception that is blocking their progress. As you monitor your students' development of rich understanding, you can continually compare their responses with your expectations and then adapt your instructional plans to accommodate their current levels of development. Wiliam (2007, p. 1054) described this interaction between teacher expectations and student performance in the following way:

> It is therefore about assessment functioning as a bridge between teaching and learning, helping teachers collect evidence about student achievement in order to adjust instruction to better meet student learning needs, in real time.

Wiliam notes that for teachers to get the best information about student understandings, they have to know how to facilitate substantive class discussions, choose tasks that include opportunities for students to demonstrate their learning, and employ robust and effective questioning strategies. From these strategies, you must then interpret student responses and scaffold their learning to help them progress to more complex ideas.

Characteristics of Tasks

The type of task that is presented to students is very important. Tasks that focus only on procedural aspects may not help students learn a mathematical idea deeply.

Superficial learning may result in students forgetting easily, requiring reteaching and potentially affecting how they understand mathematical ideas that they encounter in the future. Thus, the tasks selected for inclusion in this volume emphasize deep learning of significant mathematical ideas. These rich, "high-quality" tasks have the power to create a foundation for more sophisticated ideas and support an understanding that goes beyond "how" to "why." Figure 0.2 identifies the characteristics of a high-quality task.

As you move through this volume, you will notice that it sequences tasks for each mathematical idea so that they provide a cohesive and connected approach to the identified concept. The tasks build on one another to ensure that each student's thinking becomes increasingly sophisticated, progressing from a novice's view of the content to a perspective that is closer to that of an expert. We hope that you will find the tasks useful in your own classes.

A high-quality task has the following characteristics:
Aligns with relevant mathematics content standard(s)
Encourages the use of multiple representations
Provides opportunities for students to develop and demonstrate the mathematical practices
Involves students in an inquiry-oriented or exploratory approach
Allows entry to the mathematics at a low level (all students can begin the task) but also has a high ceiling (some students can extend the activity to higher-level activities)
Connects previous knowledge to new learning
Allows for multiple solution approaches and strategies
Engages students in explaining the meaning of the result
Includes a relevant and interesting context

Fig. 0.2. Characteristics of a high-quality task

Types of Questions

The questions that you pose to your students in conjunction with a high-quality task may at times cause them to confront ideas that are at variance with or directly contradictory to their own beliefs. The state of mind that students then find themselves in is called *cognitive dissonance*, which is not a comfortable state for students—or, on occasion, for the teacher. The tasks in this book are structured in a way that forces students to deal with two conflicting ideas. However, it is through the process of negotiating the contradictions that students come to know the content much more deeply. How the teacher handles this negotiation determines student learning.

You can pose three types of questions to support your students' process of working with and sorting out conflicting ideas. These questions are characterized by their potential to encourage reversibility, flexibility, and generalization in students' thinking (Dougherty 2001). All three types of questions require more than a one-word or one-number answer. Reversibility questions are those that have the capacity to change the direction of students' thinking. They often give students the solution and require them to create the corresponding problem. A flexibility question can be one of two types: it can ask students to solve a problem in more than one way, or it can ask them to compare and contrast two or more problems or determine the relationship between or among concepts and skills. Generalization questions also come in two types: they ask students to look at multiple examples or cases and find a pattern or make observations, or they ask them to create a specific example of a rule, conjecture, or pattern. Figure 0.3 provides examples of reversibility, flexibility, and generalization questions related to statistics.

Type of question	Example
Reversibility question	If the mode is 15, the median is 17, and mean is 20 for a set of 10 data points, what data points could be in the data set?
Flexibility question	How many different ways can you find the mean of this sample?
Flexibility question	Given these statistics for two samples, in what ways can you say that the samples are alike? In what ways are they different?
Generalization question	If a data point is added to a data set, what are some likely effects on the mean?
Generalization question	Why might using a p-value be better than using an α-value and rejection regions in hypothesis testing?

Fig. 0.3. Examples of reversibility, flexibility, and generalization questions

Conclusion

The Introduction has provided a brief overview of the nature of—and necessity for—pedagogical content knowledge. This knowledge, which you use in your classroom every day, is the indispensable medium through which you transmit your understanding of the big ideas of the mathematics to your students. It determines your selection of appropriate, high-quality tasks and enables you to ask the types of questions that will not only move your students forward in their understanding but also allow you to determine the depth of that understanding.

The chapters that follow describe important ideas related to learners, curricular goals, instructional strategies, and assessment that can assist you in transforming your students' knowledge into formal mathematical ideas related to statistics. These chapters provide specific examples of mathematical tasks and student thinking for you to analyze to develop your pedagogical content knowledge for teaching statistics in grades 9–12 or to give you ideas to help other colleagues develop this knowledge. You will also see how to bring together and interweave your knowledge of learners, curriculum, instructional strategies, and assessment to support your students in grasping the big ideas and essential understandings and using them to build more sophisticated knowledge.

Students in grades 9–12 have already had some experiences that affect their initial understanding of statistics. Furthermore, they have developed some ideas about statistics at earlier levels. Students in middle-grades classrooms frequently demonstrate understanding of mathematical ideas related to statistics in a particular context or in connection with related topics. Yet, in other situations, these same students do not demonstrate that same understanding. As their teacher, you must understand the ideas that they have developed about statistics in their prior experiences so you can extend this knowledge and see whether or how it differs from the formal mathematical knowledge that they need to be successful in reasoning with or applying statistics. You have the important responsibility of assessing their current knowledge related to the big ideas of statistics as well as their understanding of various representations of these ideas and their power and limitations. Your understanding will facilitate and reinforce your instructional decisions. Teaching the big mathematical ideas and helping students develop essential understandings related to statistics is obviously a very challenging and complex task.

into practice

Chapter 1
Statistical Modeling

Big Idea 1
Data consist of structure and variability.

> Essential Understanding 1*a*
> Mathematical models describe structure.
>
> Essential Understanding 1*b*
> Statistical models extend mathematical models by describing variability around the structure.
>
> Essential Understanding 1*c*
> Statistical models are evaluated by how well they describe data and whether they are useful.

To paraphrase the statistician George E. P. Box, all models are inaccurate, but some are useful (Box and Draper 1987, p. 424). The first big idea and its associated essential understandings in *Developing Essential Understanding of Statistics in Grades 9–12* (Peck, Gould, and Miller 2013) relate to the importance of structure and variability in understanding a set of data. Or, in the context of Box's observation, this big idea and its essential understandings address the question, "How inaccurate can a model be and still be useful?" These concepts also warn that statistical models differ from mathematical models in significant, even though sometimes subtle, ways. Statisticians need accurate models of data to make inferences, including predictions, about the population under study. The importance of accurate statistical models is reflected in the following high school standard from the Common Core State Standards for Mathematics (CCSSM; National Governors Association Center for Best Practices and Council of Chief State School Offices [NGA Center and CCSSO] 2010):

Understand and evaluate random processes underlying statistical experiments

2. Decide if a specified model is consistent with results from a given data-generating process, e.g., using simulation. (S-IC.2, p. 81)

What Constitutes a Statistical Model?

A statistical model can be as simple as pairing two variables by noticing that the value of one variable seems to be related to the value of the second variable. As an example, the table in figure 1.1 shows the number of oil changes per year for several cars, along with the associated cost of engine repairs. A model for this data might associate the annual cost of repairs for a car with the number of oil changes that the car has each year. In this case, the number of oil changes that the car has each year is the *predictor variable*, and the annual cost of repairs is the *response variable*.

Oil changes per year	X_1	3	5	2	3	1	4	6	4	3	2	0	10	7
Annual cost of repairs ($)	Y	300	300	500	400	700	400	100	250	450	650	600	0	150

Fig. 1.1. The number of oil changes per year and the annual cost of engine repairs for each of thirteen cars. From Illuminations, http://illuminations.nctm.org/Lesson.aspx?id=1189.

By contrast, someone else might believe that the annual cost of repairs for a car is more closely associated with the age of the car than with the number of oil changes that the car has each year. A model for this hypothesis might use the same response variable as the previous one (annual cost of repairs) but consider the car's age as the predictor variable. The table in figure 1.2 shows this pairing of variables.

Age of car (years)	X_2	4	3	5	5	6	4	3	3	6	8	3	2	5
Annual cost of repairs ($)	Y	300	300	500	400	700	400	100	250	450	650	600	0	150

Fig. 1.2. The age and annual repair cost for each of thirteen cars

A third person might believe that the best model would take *both* of these possible predictor variables into consideration, hypothesizing that a car's annual cost of repairs is associated with both its age and the number of oil changes that it has per year. A model for this hypothesis would have the same response variable as before (annual cost of repairs) but two predictor variables (number of oil changes per year and age of the car). The table in figure 1.3 displays this information for each car.

Oil changes per year	X_1	3	5	2	3	1	4	6	4	3	2	0	10	7
Age of car (years)	X_2	4	3	5	5	6	4	3	3	6	8	3	2	5
Annual cost of repairs ($)	Y	300	300	500	400	700	400	100	250	450	650	600	0	150

Fig. 1.3. Oil change, age, and annual repair cost for each of thirteen cars

Suppose that students were considering these three models. The question that they would want to address is, "Which of these models provides the best explanation of the difference in annual repair costs from car to car?" Knowing to ask this question and being able to answer it are at the heart of Essential Understanding 1c: "Statistical models are evaluated by how well they describe data and whether they are useful."

Inherent in understanding statistical models is a conceptual grounding in the idea of *variability*. The data in figures 1.1, 1.2, and 1.3 offer an opportunity to take a brief look at variability, which is the focus of Chapter 2. Figure 1.4 represents the data in the table in figure 1.1 in a scatterplot. A quick glance at the scatterplot would seem to support the idea that a relationship exists between the number of oil changes that a car has each year and its annual repair costs, and that a model describing this relationship could be constructed. However, your students might have difficulty verbalizing exactly what this means, since this relationship isn't a function in the mathematical sense.

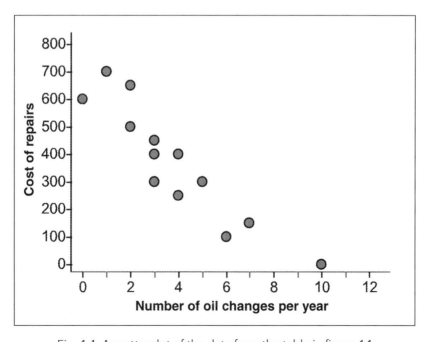

Fig. 1.4. A scatterplot of the data from the table in figure 1.1

In inspecting the table and the scatterplot, your students might note that some values of the predictor variable have unique corresponding values for the response variable—for example, (1, 700) and (10, 0). However, other values of the predictor variable have multiple corresponding values for the response variable—for instance, (3, 300), (3, 400), and (3, 450). So you might usefully ask your students to discuss precisely what it means to say that the annual cost of a car's repairs is associated with the number of oil changes that the car has per year. Reflect 1.1 invites you to consider how your students might approach this question. Pause and take time to think about your response to the questions posed in Reflect 1.1 (and, in their turn, all the subsequent questions for reflection that you encounter as you read the book).

Reflect 1.1

Suppose that your students had the table in figure 1.1 and the scatterplot in figure 1.4. How do you think they would respond if you asked, "What does it mean to say that the annual cost of a car's repairs is related to the number of oil changes that the car has per year?"

To interpret this model properly, students would need to realize that the data on these thirteen cars compose just a small sample of the data from all the cars that could have been included in the investigation. Furthermore, the underlying relationship under discussion concerns the *average* annual cost of repairs for all cars in the population from which the sample was drawn that have an equal number of oil changes per year. This is where the critically important idea of variability comes into play. For a given value X_1 of the predictor variable (number of oil changes per year), each observed value Y of the response variable (annual cost of repairs) can be expressed as $Y = \mu_1 + \varepsilon_1$, where μ_1 is the average annual cost of repairs for cars that have X_1 oil changes per year, and ε_1 measures how much Y, the annual cost of repairs for a selected car, varies from the average annual cost of repairs for all cars that have an equal number of oil changes per year. In general, the closer each ε_1 is to zero, the stronger the relationship is between X_1 and Y, and hence the more useful the model is in predicting annual cost of repairs.

A similar model could be used to describe the relationship between the variables in the table in figure 1.2. Figure 1.5 represents the data in that table in a scatterplot. In this case, the model can be expressed as $Y = \mu_2 + \varepsilon_2$, where μ_2 is the average annual cost of repairs for cars that are X_2 years old, and ε_2 is a measure of how much each Y_i differs from μ_2. Examine the questions in Reflect 1.2 to compare the usefulness of this model and the previous one for predicting a car's annual repair costs and to predict your students' choice.

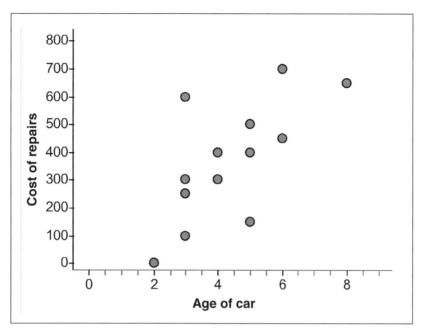

Fig. 1.5. A scatterplot of the data from the table in figure 1.2

Reflect 1.2

On the basis of the scatterplots in figures 1.4 and 1.5, which model do you think would be better for predicting a car's annual repair cost—the model using number of oil changes per year or the model using age of car?

How do you think your students would answer this question?

A linear relationship between X_1 and Y (or X_2 and Y) would be a specific, quantitative way of modeling the observed relationship in the scatterplot in figure 1.4 (or fig. 1.5). Least-squares techniques can be used to find the line of best fit for each of the scatterplots. The equation of the least-squares line in figure 1.6 is $\hat{Y} = 650.0 - 73.1X_1$, and the equation of the least-squares line in figure 1.7 is $\hat{Y} = -9.1 + 86.3X_2$. Students should be able to see that both of these models are "inaccurate" insofar as the fitted lines do not pass through each of the data points. However, the first model appears to be more "useful" since there is less variation in the points about the fitted regression line in figure 1.6 than there is in figure 1.7.

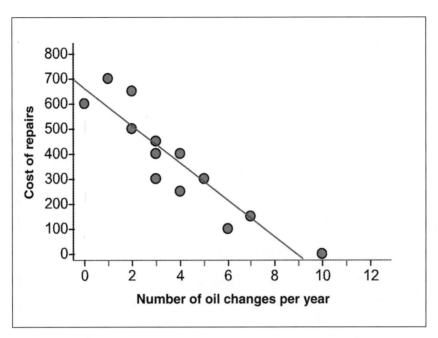

Fig. 1.6. The scatterplot from figure 1.4 along with the least-squares line

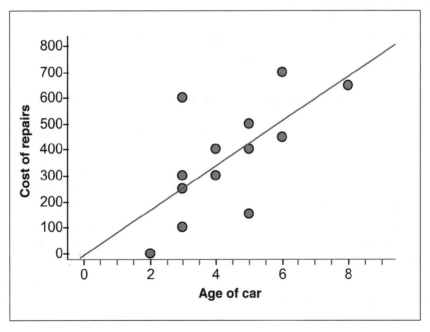

Fig. 1.7. The scatterplot from figure 1.5 along with the least-squares line

Finally, a linear model using both X_1 and X_2 can be built from the data in the table in figure 1.3; the least-squares line is $\hat{Y} = 442.2 - 61.62X_1 + 37.42X_2$. However, since the model has more than one predictor variable, this line cannot be displayed in a two-dimensional graph. Linear models with more than one predictor variable are beyond the scope of this book and will not be considered further.

Garfield and Ben-Zvi (2007) note that students tend to see a data set as individual values (each with its own characteristics) and not as an aggregate (a group with properties that may not be possessed by any individual member). Citing Hancock, Kaput, and Goldsmith (1992, p. 355), they go on to say, "To be able to think about the data as an aggregate, the aggregate must be constructed by the student" (p. 18).

In describing research efforts to help students make the conceptual leap between these two ways of viewing data, Ben-Zvi, Garfield, and Zieffler (2006) use the terms *local understanding* and *global understanding* of data and describe them as follows:

- Local understanding of data (or individual-based reasoning) involves focusing on an individual value or a few of them within a group of data (a particular entry in a table of data, a single point on a graph).

- Global understanding (or aggregate-based reasoning) refers to the ability to search for, recognize, describe, and explain general patterns in a set of data (change over time, trends) by naked-eye observation of distributions and by means of statistical parameters or techniques. (p. 470)

They further state, "Identifying patterns in a statistical graph depends on seeing the data set as a whole, taking into account the variability within the data, and integrating individual-based reasoning in some situations" (p. 472). They conjecture that the difficulty of making the transition from thinking about individual cases to aggregate-based reasoning may result from the differences between mathematical and statistical reasoning.

In working with a mathematical model for a relationship between variables, students come to expect that each point in a graph tells something about the deterministic relationship between variables, whereas in working with a statistical relationship, they must learn that the variation inherent in data means that being too focused on individual points' departures from a general trend can obscure their view of the general trend. Reflect 1.3 offers an opportunity to explore this difference with respect to the scatterplot in figure 1.8.

Reflect 1.3

Consider the data presented in figure 1.8 on U.S. coal production over the past six decades. Suppose that one student looks at the data from a local perspective and another looks at them from a global perspective.

How might the observations of the student with a local understanding differ from those of the student with a global view?

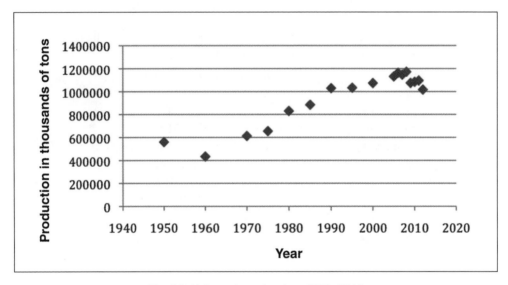

Fig. 1.8. U.S. coal production, 1950–2012

A student who exhibits only a local understanding of the data in figure 1.8 might notice that U.S. coal production was around 400 million tons in 1960 or that current annual production is at about 1 billion tons. A student with a global understanding of these same data might comment that U.S. coal production appears to have dipped between 1950 and 1960, generally increased from 1960 to about 2005, and has been decreasing since that time. Although a local understanding of data is often useful, the study of statistics is about *both* local *and* global understanding of data. As a result, students need to have a sufficient number of classroom experiences to allow them to develop a global understanding of data as well as a local understanding. Global understanding includes the ability to spot general trends and to see past the variation of individual points. However, students still must retain local understanding so that they can recognize when individual points depart from a general trend.

Statistics versus Mathematics

Before continuing, we emphasize an important point: Students can benefit from understanding the differences between *statistics* and *mathematics*. Rossman, Chance, and Medina (2006) describe several of these differences: "In statistics, context [of a problem] is crucial...[whereas] mathematicians often strive to strip away the context that can get in the way of studying the underlying structure [of a problem]" (p. 323). They also remark, "Another important issue that distinguishes statistics from mathematics is that measurement issues play a large role in statistics" (p. 326). Moreover, in statistics, the data needed are often difficult to measure.

Issues of how to collect data and how much data to collect are critical in statistics but do not often feature in mathematical problem solving. Another distinction is the lack of definitive conclusions that sometimes characterizes work on a statistical problem. When students are first exposed to statistical thinking, they may incorrectly equate a lack of definitive conclusions with the nonexistence of a correct answer.

Cobb and Moore (1997) flesh out the different values that mathematicians and statisticians give to context:

> In mathematics, context obscures structure. Like mathematicians, data analysts also look for patterns, but ultimately, in data analysis, whether the patterns have meaning, and whether they have any value, depends on how the threads of those patterns interweave with the complementary threads of the story line. In data analysis, context provides meaning. (p. 803)

Developing students' statistical thinking skills includes developing their understanding of this important distinction. Statistics and mathematics differ when it comes to the role of context. Whereas mathematicians often try to ignore context, statisticians rely on context to interpret their results.

Equally important is the difference between *statistical reasoning* and *statistical thinking*. Garfield and Ben-Zvi (2007) describe and differentiate the two. On the one hand, statistical reasoning is

> the way people reason with statistical ideas and make sense of statistical information. Statistical reasoning may involve connecting one concept to another (e.g., center and spread) or may combine ideas about data and chance. Statistical reasoning also means understanding and being able to explain statistical processes, and being able to interpret statistical results. (p. 381)

On the other hand, statistical thinking is

> a higher order of thinking than statistical reasoning. It includes the knowing how and why to use a particular method, measure, design or statistical model; deep understanding of the theories underlying statistical processes and methods; as well as understanding the constraints and limitations of statistics and statistical inference. Statistical thinking is also about understanding how statistical models are used to simulate random phenomena, understanding how data are produced to estimate probabilities, recognizing how, when, and why existing inferential tools can be used, and being able to understand and utilize the context of a problem to plan and evaluate investigations and to draw conclusions. (p. 381)

Many researchers in statistics education—for example, Wild and Pfannkuch (1999)—believe that students learn best when they are involved in statistical investigations of problems that are set in real-life contexts. In support of this idea, Wild and Pfannkuch (1999, p. 226) present a useful framework for the statistical thinking associated with data-based inquiry. They describe their framework as consisting of four dimensions in which the student operates simultaneously, as illustrated in figure 1.9. For example, students may be defining the problem and grasping system dynamics in dimension 1 ("The Investigative Cycle"), while in dimension 2 ("Types of Thinking") they are anticipating strategic problems, and in dimension 3 ("The Interrogative Cycle") they are generating models or explanations, all to satisfy their curiosity, aroused in dimension 4 ("Dispositions"), regarding the relationship between two variables.

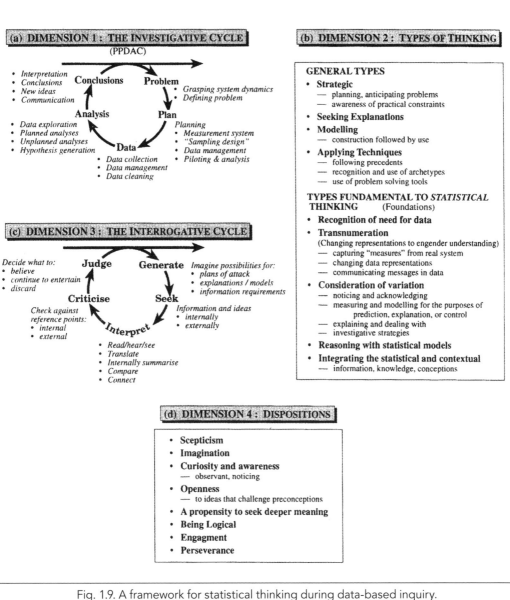

(a) DIMENSION 1 : THE INVESTIGATIVE CYCLE
(PPDAC)

- *Interpretation*
- *Conclusions*
- *New ideas*
- *Communication*

Conclusions   **Problem**
- *Grasping system dynamics*
- *Defining problem*

Analysis   **Plan**

- *Data exploration*
- *Planned analyses*
- *Unplanned analyses*
- *Hypothesis generation*

Data

Planning
- *Measurement system*
- *"Sampling design"*
- *Data management*
- *Piloting & analysis*

- *Data collection*
- *Data management*
- *Data cleaning*

(b) DIMENSION 2 : TYPES OF THINKING

GENERAL TYPES

- **Strategic**
  — planning, anticipating problems
  — awareness of practical constraints
- **Seeking Explanations**
- **Modelling**
  — construction followed by use
- **Applying Techniques**
  — following precedents
  — recognition and use of archetypes
  — use of problem solving tools

TYPES FUNDAMENTAL TO *STATISTICAL* THINKING   (Foundations)

- **Recognition of need for data**
- **Transnumeration**
  (Changing representations to engender understanding)
  — capturing "measures" from real system
  — changing data representations
  — communicating messages in data
- **Consideration of variation**
  — noticing and acknowledging
  — measuring and modelling for the purposes of prediction, explanation, or control
  — explaining and dealing with
  — investigative strategies
- **Reasoning with statistical models**
- **Integrating the statistical and contextual**
  — information, knowledge, conceptions

(c) DIMENSION 3 : THE INTERROGATIVE CYCLE

Decide what to:
- *believe*
- *continue to entertain*
- *discard*

Judge   **Generate**
Imagine possibilities for:
- *plans of attack*
- *explanations / models*
- *information requirements*

Criticise   **Seek**

Check against reference points:
- *internal*
- *external*

Interpret

Information and ideas
- *internally*
- *externally*

- *Read/hear/see*
- *Translate*
- *Internally summarise*
- *Compare*
- *Connect*

(d) DIMENSION 4 : DISPOSITIONS

- **Scepticism**
- **Imagination**
- **Curiosity and awareness**
  — observant, noticing
- **Openness**
  — to ideas that challenge preconceptions
- **A propensity to seek deeper meaning**
- **Being Logical**
- **Engagment**
- **Perseverance**

Fig. 1.9. A framework for statistical thinking during data-based inquiry.
From Wild and Pfannkuch (1999).

Wild and Pfannkuch's framework has direct and indirect connections to the big ideas and essential understandings of statistics presented by Peck, Gould, and Miller (2013) and discussed in connection with classroom practice in this book. The *problem–plan–data–analysis–conclusions* (PPDAC) investigative cycle in dimension 1 includes sampling design, data collection, and hypothesis generation, which are closely related to Big Ideas 3, 4, and 5:

> **Big Idea 3.** Hypothesis tests answer the question, "Do I think that this could have happened by chance?"

> **Big Idea 4.** The way in which data are collected matters.

> **Big Idea 5.** Evaluating an estimator involves considering bias, precision, and the sampling method.

Dimension 2 encompasses modeling and consideration of variation, with strong ties to Big Ideas 1 and 2:

> **Big Idea 1.** Data consist of structure and variability.

> **Big Idea 2.** Distributions describe variability.

The interrogative cycle *generate–seek–interpret–criticize–judge* in dimension 3 also stresses the importance of modeling and collecting data. Finally, dimension 4 includes the dispositions that students need to be successful in the framework's other three dimensions. This framework illustrates the complexity of learning statistics, as well as the various facets of statistical exploration that teachers must consider when designing statistical instruction and classroom activities. (Appendix 1 lists all the big ideas and essential understandings set out for statistics by Peck, Gould, and Miller [2013] for the reader's convenience.)

Common tenets that run throughout this book and help shape our discussions and choices of activities are the following "Principles of Learning Statistics" given by Garfield (1995, pp. 30–32):

- Students learn by constructing knowledge.

- Students learn by active involvement in learning activities.

- Students learn to do well only what they practice doing.

- Teachers should not underestimate the difficulty students have in understanding basic concepts of probability and statistics.

- Teachers often overestimate how well their students understand basic concepts.

- Learning is enhanced by having students become aware of and confront their misconceptions.

- Calculators and computers should be used to help students visualize and explore data, not just to follow algorithms to predetermined ends.

- Students learn better if they receive consistent and helpful feedback on their performance.

- Students learn to value what they know will be assessed.

- Use of the suggested methods of teaching will not ensure that all students will learn the material.

These principles provide valuable guidance to teachers as they consider how to incorporate statistical investigations into their teaching practices. Of particular importance are the principle that having students become aware of and confront their misconceptions enhances learning and the related idea, implicit in many principles in the list, that students should experience some degree of struggle as they learn. Reflect 1.4 asks you to think about this second idea.

Reflect 1.4

Do you agree that students should struggle somewhat as they learn?

When does struggle help them learn?

Under what circumstances might struggle inhibit their learning?

Statistics should not be taught as if it is in its final form, consisting of formulas, worked examples, and memorized theorems that are transmitted to students. Statistical learning is active and dynamic and can advance when students make mistakes. Students are defining problems, making conjectures, constructing plans for data collection, exploring data and testing conjectures, and estimating parameters while concurrently interpreting and critiquing their results with skepticism, perseverance, and a willingness to refine their processes. Seeley (2009, p. 2) offers a useful observation about the value of this activity:

> Constructive struggling can happen when a skillful teacher gives students engaging yet challenging problems. Constructive struggling can take place when a teacher decides that one demanding, possibly time-consuming problem will likely provide more learning value than several shorter but more obvious problems. Constructive struggling involves presenting students with problems that call for more than a superficial application of a rote procedure.... As students engage in the constructive struggling needed for some of these problems, they learn that perseverance, in-depth analysis, and critical thinking are valued in mathematics as much as quick recall, direct skill application, and instant intuition.

However, the skillful teacher knows when to intervene so that constructive struggling does not reach the level of frustration that causes students to give up and quit trying to solve the problem.

A Statistics Scenario: Recycling at Red Mountain High School

Throughout this book, we will be using a scenario about a student-sponsored recycling program to illustrate various key points related to the teaching and learning of statistics. The scenario moves from the project's formation through the students' collecting of data, testing of hypotheses, predicting of parameters, and forming of conclusions. The scenario begins by introducing the primary participants, Morgan and Mackenzie, two high school students who launch a data investigation.

> Morgan and Mackenzie are members of the Ecology Club at Red Mountain High School (RMHS), and they are concerned about the amount of trash they see in the trash cans in the RMHS classrooms. They are determined to help reduce the amount of trash that winds up in the city's landfill but are unsure exactly what to do. At the next meeting of the Ecology Club, Mackenzie raises a question about putting recycling bins in each of the school's classrooms. Morgan quickly agrees that RMHS needs a recycling program and suggests that the Ecology Club approach the principal, Ms. Begay, about buying recycling bins. The members of the Ecology Club unanimously vote yes—Red Mountain High School needs to begin a schoolwide recycling program.
>
> "How many recycling bins will the school need to purchase?" asks Mackenzie.
>
> Antonio, the club's president, replies, "Well, there are fifty classrooms in the main building, but I think we should also put bins in the gym, both the boys' and girls' locker rooms, the library, the main office, and the faculty lounge. I can think of several other places too. I bet we could use at least seventy-five recycling bins."

Mr. Diaz, the club's faculty sponsor, holds up a hand. "Whoa," he says. "That's great, but I think you're moving too fast. It would be very expensive to buy seventy-five recycling bins, and Ms. Begay is never going to agree to spend that much money. But she might let you put recycling bins in some of the rooms if you can convince her that the students, faculty, and staff will use them."

"Mr. Diaz, how can we convince her?" asks Morgan.

"You'll need to use statistics to make a persuasive argument about why it would be a good idea to get recycling bins and start a recycling program," Mr. Diaz replies.

"That's a great idea!" says Morgan. "We can put recycling bins in some of the rooms, and then measure the amount of recycling in the bins to convince Ms. Begay that we should put bins all over the school."

Antonio has a puzzled look on his face. "But we can't compare the amount of recycling that is in the rooms with bins to the recycling in rooms without bins—there won't be any."

"That's a good point, Antonio," says Mr. Diaz.

Mackenzie asks, "Am I the only one who's confused? How do we know how many bins we should ask Ms. Begay to buy, and what rooms we should put them in? How will we know if people will use them after they're bought?"

Mackenzie's confusion has important implications and offers a good opportunity to pause and reflect. Consider the question in Reflect 1.5.

Reflect 1.5

How would you respond to Mackenzie's questions?

Notice that when Mr. Diaz responds to Mackenzie he refrains from saying too much too soon:

"Those are important questions that must be answered, Mackenzie," Mr. Diaz says. "The first thing you need to do is decide what data to collect to make a strong argument to Ms. Begay. Antonio pointed out that you can't really compare the amount of recycling between rooms with bins and rooms without. So what variables can you measure?"

"I know," says Antonio. "It doesn't make sense to measure the amount of recycling, but we can measure the amount of trash in a room's trash can."

"Good start," says Mr. Diaz. "How would you use that information?"

"Ummmm, well, wouldn't there be more trash in rooms that don't have a recycling bin?" offers Antonio.

Mr. Diaz smiles at Antonio. "Sounds as though you're talking about a statistical model."

Several members of the club simultaneously ask, "What's that?"

Mr. Diaz explains, "A model describes how one or more variables are related to one or more other variables. You could construct very sophisticated statistical models to use in your recycling project. However, in this case, I would suggest that you make your model as simple as possible. What are your variables in this situation?"

Mackenzie speaks slowly, "Well, like Antonio said, one variable is the amount of trash in each room. Do we need another one too?"

"OK," says Mr. Diaz. "So, here you have two variables that can be put together into a statistical model. One is the *predictor variable,* which is the one that you manipulate from case to case and that you think influences the outcome of the other. The other variable, the one whose value you have no control over, is called the *response variable.* In your model, what do you think is the predictor variable and what is the response variable?"

Morgan says, "I think the amount of trash in a room can depend on whether or not there is also a recycling bin in the room. So, the predictor variable is the presence of a recycling bin, and the response variable is the amount of trash."

"Excellent!" says Mr. Diaz. "Now you have the beginnings of a statistical model. Do you have any ideas what some of your model's properties might be?"

Mackenzie answers, "I think the rooms without recycling bins will have more trash than the rooms with recycling bins. Let's go talk to Ms. Begay."

Morgan chimes in. "Wait! How will we know if the amount of trash is reduced? In my second-period class, my desk is near the trash can by the door, and sometimes there's a lot of trash in it, and sometimes there's hardly any. Won't that affect our results?"

We will return to the recycling scenario in Chapter 2. But first we want to discuss how students' understanding of statistical models, and their facility in building and working with them, can be enhanced by the classroom use of *model-eliciting activities.*

Model-Eliciting Activities

Developing a statistical model is typically more intricate than constructing a mathematical equation. However, students' exposure to model-eliciting activities helps them build a robust understanding of the components of a statistical model. Larson (2010) refers to a model as a conceptual system that students use to make sense of a problem situation. She quotes Lesh and colleagues (2000, p. 609):

> A model is a system that consists of (a) *elements;* (b) *relationships among elements;* (c) *operations* that describe how the elements interact; and (d) *patterns* or *rules …* that apply to the relationships and operations. However, not all systems function as models. To be a model, a system must be used to describe another system, or to think about it, or to make sense of it, or to explain it, or to make predictions about it.

Furthermore, Carmona and Greenstein (2010) describe a model-eliciting activity as a real-life problem to be solved by small groups of students who must construct a mathematical or statistical model that adequately represents the real-world system presented in the problem and clearly describes, explains, or predicts the behavior of the system. Further, although model-eliciting activities typically do not lead to a single solution, there are usually optimal ways to solve the problem represented by the model. Hamilton and colleagues (2008) comment that students' "iterations through the 'express-test-revise' cycle of model revision can yield new cognitive structures and understandings … more effectively than a single iteration application of [a] textbook formula," and they observe that the "solution orientation" of these activities "enables crucial development of complex reasoning processes, and suggests an alternative balance for how 'product' and 'process' are emphasized in the curriculum" (p. 4).

Lesh, Amit, and Schorr (1997) offer guidance about constructing model-eliciting activities by describing six fundamental characteristics that such activities possess when they are well designed. Well-crafted model-eliciting activities embody the following principles:

- *The Reality Principle:*

  The context of the problem should be set in a real-life situation. The students must be allowed to use their own personal knowledge and experiences to make sense of the problem instead of following the teacher's "correct" way of thinking.

- *The Model Construction Principle:*

  The students must create a model to solve the problem. Furthermore, the model is constructed by focusing on underlying rather than superficial characteristics.

- *The Self-Evaluation Principle:*

  Students are given clear criteria so that they can judge for themselves whether their model is appropriate.

- *The Model-Documentation Principle:*

  The model-eliciting activity must require students to explicitly communicate (orally and/or in writing) the thinking processes used while solving the problem.

- *The Model Generalization Principle:*

  The model that is constructed should be able to be applied to a broad range of situations. It should present a general way of thinking rather than a specific way of solving the specific given problem.

- *The Simple Prototype Principle:*

  The model should be as simple as possible, but also significant. The solution should provide a useful prototype for interpreting a variety of other structurally similar situations. (pp. 2–3)

Model-eliciting activities are different in structure and purpose from the typical activities that teachers select and use. However, they are not meant to replace regular classroom activities but rather to be offered in conjunction with them. The website *Pedagogy in Action: The SERC Portal for Educators* presents a table, reproduced in figure 1.10, that describes how these differences may appear in a statistics classroom.

Regular activities	Model-eliciting activities (MEAs)
One or more may be used in each class session.	Only a few may be used throughout an entire course.
Are used to accomplish a particular learning goal, usually to illustrate a concept or procedure or to discover a statistics idea (e.g., having students understand why random samples are better than judgmental ones, or what makes a correlation larger or smaller).	Are used to engage students in thinking and reasoning that will set the foundations for *future learning of a concept or procedure*. The goal is to develop a model to use in solving a problem, but students are not expected to develop what might be "the right" model.
Lead students to a desired solution or conclusion.	Are designed to lead to many different conclusions or solutions that need to be explained and justified by students.
May or may not be used to promote classroom discourse.	Promote classroom discourse by having students discuss and critique various solutions.
Students either produce or learn the "correct" solution to a problem or question.	Students' solutions are compared, but no correct solution is given. However, an expert solution may be provided at a later time, for comparison.
It is often clear what steps students should follow in the activity (e.g., use this applet, take this sample of data).	The problem is open-ended and steps are not given for students to follow; however, the needs of a client and the data provided allow students to determine when their solution is complete.
The activity may or may not be related to a real problem or real data.	The problem should be real and based on real data (or partially modified real data).
May be completed by students working alone.	Students initially read the problem and respond to some questions alone, but then work together as a team to produce and explain their model.
Reading and writing may or may not be a component of the activity.	Reading background information and writing a letter or report is part of the MEA.

Fig. 1.10. A table comparing model-eliciting activities with typical in-class activities.
From *Pedagogy in Action: The SERC Portal for Educators,*
http://serc.carleton.edu/sp/library/mea/what.html.

Using a Model-Eliciting Activity: iPod Shuffle

Figure 1.11 presents an example of a model eliciting activity: the iPod Shuffle problem from delMas, Garfield, and Zieffler (2009). An iPod owner, Albert Hoffman, has written a letter to Apple complaining that the shuffle feature on his iPod is not generating random playlists. The activity is designed for students to work on in teams that are given sets of randomly generated playlists that they can use to devise, test, and revise models to determine whether or not the iPod playlists are actually generated randomly. (Before giving the activity to your students, you might want to substitute the name of the current Apple CEO, Tim Cook, in place of the name of deceased Apple cofounder and CEO Steve Jobs.)

Questions to Get You Started

1. Do you have an iPod or some other digital music player? Have you used the shuffle feature? If you have used the shuffle feature, have you ever wondered how truly random it is?

2. What comes to mind when you hear the word "random"?

3. If the iPod shuffle feature is *not* producing a random sequence of songs, then what might the sequence of songs look like? What would you expect to see?

4. Do you think you can be 100% certain that a sequence of songs was not randomly generated? Explain your answer.

Once everyone in your group has answered all four questions, share and discuss your responses to each question.

Group Task

Albert Hoffman, an iPod owner, has written a letter to Apple to complain about the iPod shuffle feature. He writes that every day he takes an hour-long walk and listens to his iPod, using the shuffle feature. **He believes that the shuffle feature is producing playlists in which *some artists are played too often and others are not played enough*.**

He has claimed that the iPod Shuffle feature is not generating random playlists. As evidence, Mr. Hoffman has provided both his music library (8 artists with 10 songs each) and three playlists (20 songs each) that his iPod generated, using the shuffle feature.

Steve Jobs, the CEO of Apple, Inc., has contacted your group to respond to Mr. Hoffman's complaint. He has provided your group with several playlists of 20 songs, each using the same songs as Mr. Hoffman's library but generating them by using a genuine random number generation method.

To help your group respond to Mr. Hoffman, the next four parts of the problem are designed to help your group explore properties of the randomly generated lists to develop rules that could help determine whether a set of playlists provide evidence that the shuffle feature is not producing randomly selected songs.

Part I: Explore and describe
Examine the randomly generated playlists (your group will be given 25) to get an idea of the characteristics of these lists. Write down and number two or more characteristics of a randomly generated playlist.

Part II: Develop rules
Use the set of characteristics that your group wrote down to describe randomly generated playlists in part I to create a set of one or more rules that flag playlists that **do not appear to have been randomly generated.** (Be sure that each of the characteristics in Part I is included in a rule.) These rules should be clearly stated so that another person could easily use them.

Part III: Try out rules
Your group will be given five additional randomly generated playlists on which to test your rules. See whether the set of rules your group generated would lead someone to (incorrectly) question whether these playlists are not randomly generated. Based on the performance of your group's set of rules, adapt or change the rules as your group feels necessary.

Part IV: Evaluate
Your group will be provided with Mr. Hoffman's original three playlists. Apply your group's rules to these three playlists to judge whether there is convincing evidence that Mr. Hoffman's iPod Shuffle feature is producing playlists that do not seem to be randomly generated.

Part V: Summarize
Your group will now write a letter to Mr. Hoffman that includes the following:

1. Your group's set of rules used to judge whether a playlist does not appear to have been randomly generated. In your letter the rules need to be clearly stated so that another person could apply them to a playlist of 20 songs from Mr. Hoffman's music library;

2. A response to Mr. Hoffman's claim **that the shuffle feature is not random because it produces playlists in which *some artists are played too often and others are not played enough*.**

Fig. 1.11. The iPod Shuffle activity. From delMas, Garfield, and Zieffler (2009); also available online at http://www.amstat.org/education/stew/pdfs/HowRandomIsTheiPodsShuffle.pdf.

The Pitching Machine activity in Chapter 2 and the activity Raffling to Support Recycling in Chapter 4 are other examples of model-eliciting activities. Additional samples can be found at a variety of websites, including serc.carleton.edu/sp /library/mea/examples.html or http://www.cpalms.org/cpalms/mea.aspx.

Model-eliciting activities can help students develop statistical thinking by engaging them in problem-solving situations in which they consider data production, operationalize constructs, test and refine models, and justify their conclusions (Garfield, delMas, and Zieffler 2010). In turn, this improvement in statistical thinking can strengthen their ability to build statistical models, as well as their development of understanding and skill in using the other big ideas and essential understandings identified and addressed in subsequent chapters.

Conclusion

Chapter 1 has introduced a scenario about students' development of a high school recycling project that subsequent chapters will extend to demonstrate some of the ideas that the book explores. When students actively analyze and model data, they learn that a statistical model is often as simple as an association between a predictor variable and a response variable. Gradually, they learn that the value of a statistical model rests on its usefulness for making predictions or inferences about a population, and they see that the model's usefulness depends in turn on the context of the data as well as the variability that the data demonstrate under the particular model. In general, the less variability, the more useful the model can be for making predictions or inferences.

This chapter has explored students' tendency to focus on individual data values rather than aggregated data, perhaps because of the students' greater familiarity with mathematical models, in which all points possess the same properties. Activities that help students develop a global perspective while retaining a local perspective on data can foster their statistical thinking. Further, because students' learning is robust when they construct knowledge actively with some degree of struggle, model-eliciting activities can be very helpful.

Chapter 2 focuses on ways to help students grasp the fundamental idea of variability in data, introduced briefly in Chapter 1, and it extends the scenario about the recycling project at Red Mountain High School to develop ideas related to this challenging concept.

Chapter 2
Variability

Big Idea 2
Distributions describe variability.

Essential Understanding 2*a*
A population distribution describes variability in the values that make up a population.

Essential Understanding 2*b*
The population distribution is often unknown but can be approximated by a sample distribution.

Essential Understanding 2*c*
The sampling distribution of a sample statistic describes how the value of the statistic varies from sample to sample.

Essential Understanding 2*d*
Simulation can be used to approximate sampling distributions.

Samuel Johnson is often said to have remarked, "You don't have to eat the whole ox to know that the meat is tough." Although the story of Johnson having made this observation may be apocryphal, many who have discussed sampling have quoted the statement, which provides an apt metaphor for the use of sampling in data analysis. This is the idea at the heart of sampling: sampling and variability are inextricably linked. The second big idea and its accompanying essential understandings in *Developing Essential Understanding of Statistics for Teaching Mathematics in Grades 9–12* (Peck, Gould, and Miller 2013) are related to population distributions, sampling distributions, and the associated variability.

Variability is everywhere. Height varies from one person to another, and you have probably found that your own weight varies somewhat throughout the day. Variation occurs in the number of hours that similarly produced light bulbs will last or the number of ounces of water in bottles filled by a machine in an assembly line. In describing the role of variability in statistics, Cobb and Moore (1997) offer the following observation:

> In some circumstances, we want to find unusual individuals in an overwhelming mass of data. In others, the focus is on the variation of measurements. In yet others, we want to detect systematic effects against the background noise of individual variation. Statistics provides means for dealing with data that take into account the omnipresence of variability. (p. 801)

Rossman and Chance (2004) amplify this statement: "Much of statistics is based on the fact that while individual values vary, their pattern of variation in the aggregate is often quite predictable" (p. 1). Further, according to Wild and Pfannkuch (1999), "It is variation that makes the results of actions unpredictable. Variation is the reason why people have had to develop sophisticated statistical methods to filter out any messages in data from the surrounding noise" (pp. 235–36). Variation in data comes from different sources and is one of the most difficult concepts for students to understand. Reflect 2.1 provides an opportunity to consider possible classroom approaches to different types of variation.

Reflect 2.1

What are some sources of variation in data?

How could you model these different types of variation in your classroom?

Is *variation* the same thing as *accuracy*?

Wild and Pfannkuch (1999) provide a diagram that is helpful in delineating the different types of variation that can be found in data. Their diagram appears in figure 2.1.

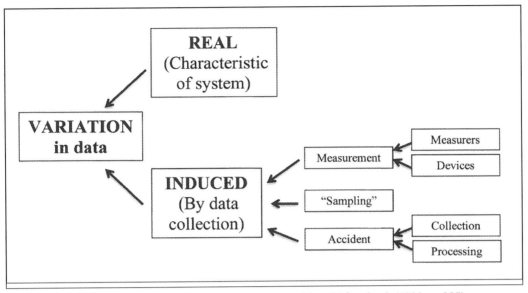

Fig. 2.1. Sources of variation in data. From Wild and Pfannkuch (1999, p. 235).

First, we need to understand the "real" variation that exists in the population and is passed on to random samples. Natural variation occurs in our heights, weights, and preferences for political candidates. We take different amounts of time to heal from injuries. Plants react differently to the amount of sunlight that they receive.

Second, we need to understand the variation in the data that make up a random sample. This variation is induced by the data collection method and may be attributed to several causes: how the data are measured, who is doing the measuring, and properties of the measuring devices. The variation may be due to the rounding of numerical values, ordinary measurement error (a lack of exact reproducibility of measurements), and so on. Sample variation may also be attributed to mistakes in collecting or processing the data—for example, miscoding, errors in data entry or analysis, and so forth. Finally, and most important, variation in data is due to the natural variation that occurs in drawing a random sample: Different samples produce different information and results, and the particular sample that we obtain is random. It is important that students realize that much of statistics is devoted to understanding variation. Students must be exposed to activities that allow them to explore all forms of variation in data so that they realize which types of variation can and which types of variation cannot be controlled.

Distinguishing *Variability* from *Accuracy* in an Inventing Activity: Pitching Machines

Students sometimes confuse *variability* with *accuracy.* They often see variability as a lack of accuracy rather than as a measure of the degree to which data deviate from their "center" (Schwartz and Martin 2004). In addition, although students may know how to compute the standard deviation of a set of data, they may not know how to explain what it means or when it should be used (Reading and Shaughnessy 2004, p. 204).

Consider the Pitching Machines example that follows, adapted from an *inventing activity* developed by Schwartz and Martin (2004). (Additional discussion of inventing activities and sample activities are available in Appendix 3.) In this activity, students are asked to devise a scheme to measure the quality (accuracy and variability) of four different pitching machines. They are given grids like those in figure 2.2, with an X that represents the middle of the target and black dots that give the locations where different pitches landed. Examine the grids, along with the instructions to students on the next page, and then respond to the question in Reflect 2.2.

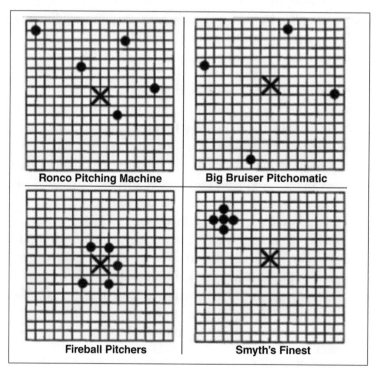

Fig. 2.2. Grids showing results for four pitching machines, with each grid representing a 16-inch by 16-inch square. Based on Schwartz and Martin (2004).

Instructions: The four grids in figure 2.2 show the results from four different pitching machines. The X represents the target, and the black dots show where different pitches landed. Your task is to invent a procedure that computes a "quality" index for determining which of the pitching machines is best. There is no single way to do this, but you have to use the same procedure for each machine so that your comparison of the machines is fair. Write your procedure and the index value that you compute for each pitching machine.

Reflect 2.2

What are some of the procedures that you might expect your students to develop to measure the quality of the pitching machines?

In the Pitching Machines activity, students develop procedures for constructing a measure of quality that incorporates both the accuracy and the variability of each machine. Further, students must consider relative weights to give to these competing measures. The best machine would have high accuracy and low variability. However, students must also consider at what point low variability might outweigh higher accuracy.

Students could measure a machine's accuracy by first measuring the distance that each black dot is from the center of the target, as shown in figure 2.3. The sum of these distances for the Ronco Pitching Machine, Big Bruiser Pitchomatic, Smyth's Finest, and Fireball Pitchers are 29.1 inches, 28.9 inches, 32.2 inches, and 11.5 inches, respectively. However, since each machine was not tested with the same number of pitches, students should realize that these sums are misleading and that an average distance is called for. These averages are 5.8 inches, 7.2 inches, 6.4 inches, and 2.3 inches, respectively. You may have to ask intervening questions to make sure that your students understand that the averages are a measure of accuracy, not variability. These values measure how close the pitches are to the target, not how close they are to one another.

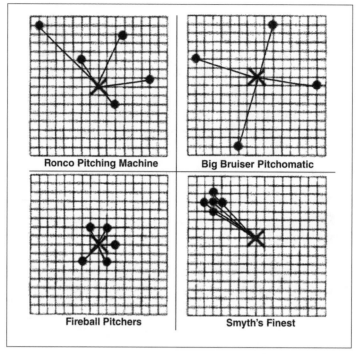

Fig. 2.3. Measuring accuracy as the distance from the center, X, to each point where a pitch landed

Next, students may suggest that an additional measure might be obtained by determining how far each of the individual distances is from the average distance, which eventually leads to the definitions of the variance and standard deviation of a set of data. The students might use several methods to compute a machine's variability, including determining the sums of the deviations and the mean absolute deviations. Furthermore, students might try to quantify variation in other ways, two of which are detailed below and illustrated in figure 2.4:

- Use the data points to form convex polygons (some students may decide to form non-convex polygons) and then measure the perimeters of those polygons. This process gives 33.6 inches for the Ronco Pitching Machine, 41.0 inches for the Big Bruiser Pitchomatic, 5.7 inches for Smyth's Finest, and 13.6 inches for the Fireball Pitchers.

- Measure the area "covered" by the data points. This gives 55.0 square inches for the Ronco Pitching Machine, 104.0 square inches for the Big Bruiser Pitchomatic, 2.0 square inches for Smyth's Finest, and 12.0 square inches for the Fireball Pitchers.

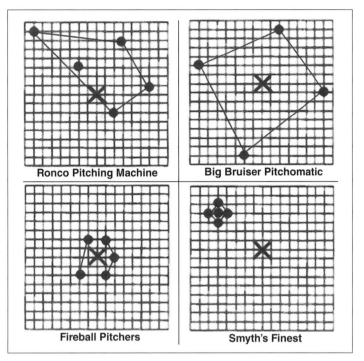

Figure 2.4. Using polygons to measure variability

After considering these two methods, pause to respond to the questions in Reflect 2.3.

Reflect 2.3

Which of the pitching machines would you (or your students) say is the most accurate?

Which is the most reliable?

How would you judge which pitching machine is "best" overall?

Students might combine measures of accuracy and variability in many ways to get an overall quality index. A simple approach might be to add the two component measures together, provided they are on the same scale. Figure 2.5 provides a table that illustrates this approach. Students might conclude that the Fireball Pitchers machine is the best. This machine has the lowest average distance from the center of the target and the second-lowest variability, as measured by either area or perimeter. Further, students might conclude that the low variability of the Smyth's Finest machine does not outweigh its inaccuracy.

Machine	Accuracy (A) (inches)	Variability 1 (V_1) (perimeter, inches)	Variability 2 (V_2) (area, square inches)	Overall quality index 1 $(A + V_1)$ (inches)	Overall quality index 2 $(A^2 + V_2)$ (square inches)
Ronco Pitching Machine	5.8	33.6	55.0	39.4	88.6
Big Bruiser Pitchomatic	7.2	41.0	104.0	48.2	155.8
Smyth's Finest	6.4	5.7	2.0	12.1	43.0
Fireball Pitchers	2.3	13.6	12.0	15.9	17.3

Fig. 2.5. Quality ratings (accuracy plus variability) of four pitching machines

However, the choice might not be so clear if Smyth's Finest were slightly more accurate. Figure 2.6 shows a different set of test results for the Smyth's Finest machine. With these results, the measures of accuracy (average distance from the center) and variability (area covered) for Smyth's Finest are 3.7 inches and 2.0 square inches, respectively, whereas they are 2.3 inches and 12.0 square inches, respectively, for the Fireball Pitchers machine, and the quality ratings change to those shown in the table in figure 2.7. Consider the last column, which shows overall quality index 2, in the tables in figures 2.5 and 2.7. An interesting classroom discussion can develop when students consider whether there is a point at which the measure of variability (or reliability) is small enough for Smyth's Finest machine that it balances its greater inaccuracy, and Smyth's Finest becomes the better choice.

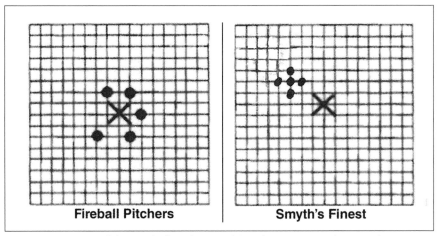

Fireball Pitchers | **Smyth's Finest**

Fig. 2.6. A comparison of the Fireball Pitchers and Smyth's Finest machines with more accurate Smyth's Finest results

Machine	Accuracy (inches)	Variability 1 (perimeter, inches)	Variability 2 (area, square inches)	Overall quality 1 (A + V1) (inches)	Overall quality index 2 (A² + V2) (square inches)
Ronco Pitching Machine	5.8	33.6	55.0	39.4	88.6
Big Bruiser Pitchomatic	7.2	41.0	104.0	48.2	155.8
Smyth's Finest	3.7	5.7	2.0	9.4	15.7
Fireball Pitchers	2.3	13.6	12.0	15.9	17.3

Fig. 2.7. A table showing quality ratings (accuracy plus variability) of four pitching machines, with improved accuracy for Smyth's Finest machine

Researchers such as Garfield and Ben-Zvi (2007) believe that statistics textbooks and instruction give too much classroom attention to measures of central tendency and too little consideration to topics related to variability. They claim that it is impossible to separate a proper understanding of a measure of central tendency from an understanding of variability. Further, variability is a complex concept to learn because of its multifaceted characteristics. Sometimes variability is of intrinsic interest and importance in an investigation, whereas at other times it is "noise" or a nuisance that interferes with understanding the "signal" in the data.

Variability not only can be seen and interpreted informally in graphs but also can be quantified and interpreted formally by using summary measures, such as the range, interquartile range, or standard deviation of a data set.

According to Shaughnessy (2006), students may develop many different conceptions of variability, including the following:

- Variability as *extremes* or *outliers*. Students conceiving of variability in this way focus their attention on the tails of a distribution of data or on strange values.

- Variability as *change over time*. Having students explore graphs involving time (e.g., wait times between eruptions of Old Faithful) is often a good first introduction to the concept of variability across repeated measurements.

- Variability as the *whole range*—the spread of all possible values.

- Variability as *the likely range* of actual measurements if an experiment were repeated many times. This conception of variability can form the basis for developing statistical tools for representing and measuring variability, like box plots and standard deviation.

- Variability as *the distance or difference from some fixed point*, such as the mean or the median. This conception of variability results in a numerical measure with a geometric interpretation.

- Variability as a *sum of* [squared] *residuals*. In this conception, variability is a measure of the total amount that a distribution is "off" from some central value. (pp. 92–93)

Reflect 2.4 offers an opportunity to think more closely about these different conceptions of variability.

Reflect 2.4

Which of the conceptions of variability identified by Shaughnessy (2006) have you seen in your classroom?

Do you think these interpretations are equally useful in conveying the concept of variability to students?

Exploring Variability: Recycling at Red Mountain High School, Continued

The recycling scenario introduced in Chapter 1 continues here to illustrate a natural context in which students can encounter, explore, and begin to develop an understanding of the concept of variability:

Morgan, Mackenzie, and Antonio are in Ms. Begay's office talking about their ideas for a schoolwide recycling program. Ms. Begay tells the trio that she really likes the idea of decreasing the amount of trash that the school generates, but she cautions that maybe their plan is a bit ambitious. "It's too expensive to put recycling bins in every room in the school, especially since we don't know whether our school community will support it," says Ms. Begay.

"Then what should we do?" asks Antonio.

Morgan says, "I have an idea. Our school has many more students and teachers than administrators and staff, so let's put recycling bins just in the classrooms."

Ms. Begay replies, "That's a good idea. Maybe a first step would be to put bins in just a few of the classrooms to see if they get used. I think we might be able to get four recycling bins from the district office. You can run a trial experiment to help you decide whether you want to go forward with the project."

"A trial experiment? Is that easy to do?" asks Mackenzie.

"There's a lot you'll have to consider," says Ms. Begay. "You'll have to think about how to determine which classrooms to include in your experiment—your sample, that is—how you'll collect the data, and what kind of analysis you'll do, among other things."

"*Sample? Data? Analysis?*" Mackenzie asks in a troubled voice.

"It's not as bad as you think," Ms. Begay replies. "I can get you started, but then you should talk to the statistics teacher, Mrs. Lee. First, think about your model and data. What are you going to be measuring?"

Antonio speaks up. "We decided that we should measure the amount of trash in the rooms that have recycle bins and compare it with the amount of trash in the rooms that don't have the bins."

"But how are we going to make the comparison?" Morgan asks. "Are we going to compare the total amount of trash in the rooms that have recycle bins with the total amount of trash in the rooms that don't have recycle bins? That wouldn't seem to be fair unless we had the same number of rooms."

"We could compare the average amount of trash in each of the two types of rooms," suggests Antonio.

"That makes sense," says Mackenzie. "But there are a lot of classrooms in the building. I don't think we can use all of them. How many rooms should we use?"

Morgan suggests that the bins should definitely go in the computer lab and art room, since they generate a lot of paper. She also advises against putting bins in any of the language rooms since students primarily use the computer language-learning stations and rarely use any paper.

"Somehow that doesn't seem quite right to me," says Ms. Begay. "Don't you think your plan might be biased and skew the results?"

Antonio asks, "What do you mean by 'biased'?"

Mackenzie says, "I think it might be time to go talk to Mrs. Lee."

Variability and Sampling

In their research on statistical reasoning, Watson and Moritz (2000) and Sotos and colleagues (2007) highlight what is often called the "law of small numbers," which is a judgment bias caused by students' beliefs that a sample of any size will have the same characteristics as the population from which it is drawn. Examples of this type of bias include thinking such as the following:

- Fifty percent of the people in my state own some type of firearm. If I do not own a firearm, then my neighbors on either side do.

- A certain drug is effective in 75% of patients. So, if you treat four patients, three of them will respond positively.

Sotos and colleagues (2007) also say that students' belief in the law of small numbers can cause them to confuse the population distribution and the sampling distribution, as well as the difference between the distribution of a sample and the sampling distribution of a statistic. It can also cause students to underestimate the size of confidence intervals, overestimate the significance in tests of hypotheses, and be overconfident about obtaining the same results in future replications of an experiment. Many misconceptions that students harbor in relation to the law of small numbers can be traced to their misunderstanding of how the variability of samples is related to the variability in populations. Reflect 2.5 asks you to consider your students' misconceptions about this relationship.

Reflect 2.5

What are some of the sampling misconceptions that you have seen among your students?

Do your students commonly apply the "law of small numbers" in connection with sampling and the variability of samples?

How can you help your students overcome these misconceptions?

Sampling Distributions

Chance, delMas, and Garfield (2004, p. 302) identify some common misconceptions that students have about sampling:

- Sampling distributions should look like the population (for sample size n > 1).

- Sampling distributions should look more like the population as the sample size increases.

- Sampling distributions for small and large sample sizes have the same variability.

- Sampling distributions for large samples have more variability than sampling distributions for small samples.

Students also tend to—

- fail to understand that a sampling distribution is the distribution of a sample statistic;

- confuse one sample (real data) with all possible samples (in the sampling distribution) or potential samples;

- pay attention to the wrong things—for example, heights of histogram bars (when examining sample variability);

- think that the population mean of a positively skewed distribution will be greater than the mean of the sampling distribution of $\bar{x}$ for samples taken from this population.

Technology can be used to help students see through these misconceptions. Below are some screenshots from the data software Fathom. Figure 2.8 shows the graph of, and associated descriptive statistics for, a set of data consisting of 50 randomly chosen whole numbers between 0 and 20. In relation to the recycling scenario, the graph and statistics could represent the distribution of the amount of trash, measured in pounds, in the 50 classrooms of Red Mountain High School on a given day.

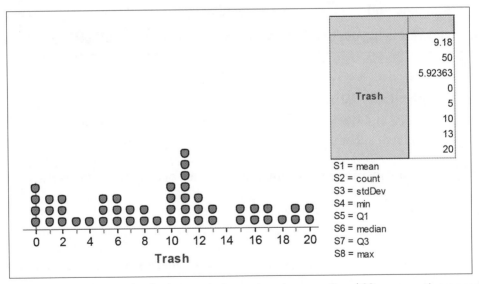

Fig. 2.8. A dot plot of 50 randomly chosen whole numbers between 0 and 20, representing pounds of trash in the cans in the 50 classrooms of Red Mountain High School on a given day

Figure 2.9 shows the graph of and associated descriptive statistics for a set of data consisting of the means of 100 samples of size 5, taken from the population in figure 2.8. Finally, figures 2.10 and 2.11 are graphs of and associated descriptive statistics for sets of data consisting of the means of 100 samples of sizes 10 and 20, respectively, taken from the population in figure 2.8.

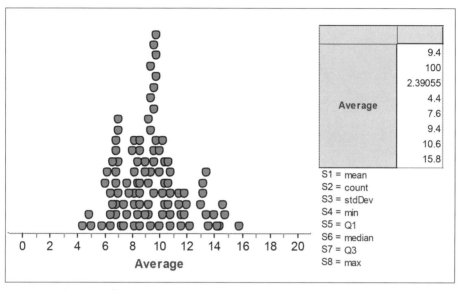

Fig. 2.9. A dot plot of 100 sample means, sample size n = 5, taken from the population shown in figure 2.8

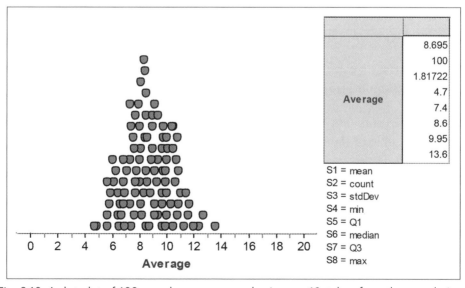

Fig. 2.10. A dot plot of 100 sample means, sample size n = 10, taken from the population shown in figure 2.8

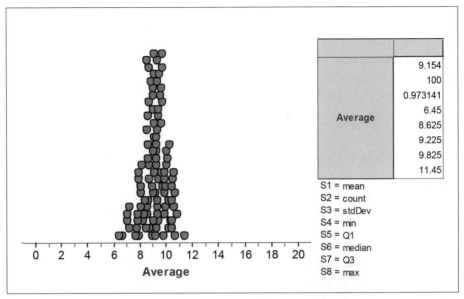

Fig. 2.11. A dot plot of 100 sample means, sample size *n* = 20, taken from the population shown in figure 2.8

These results clearly provide a counterexample to many of the misconceptions in the earlier list. In particular, they demonstrate that—

- for a fixed sample size, the distribution of sample means does not resemble the population distribution, and the lack of similarity is more striking as the sample sizes increases.

- the variability of the sampling distribution of the mean generally decreases as the sample size increases.

The key idea for students to understand is that when a random sample is properly selected, some of the characteristics of the sample will be similar to those of the population. For example, the mean of the sample is likely to be similar to the mean of the population insofar as the distribution of all possible sample means is centered at the mean of the population. But not all different samples will look the same. They will show differences from the population as a result of sampling variability. Furthermore, as Essential Understanding 2*d* recognizes, simulation is a useful tool for approximating sampling distributions and providing a better understanding of what similarities and differences to expect among populations, samples, and sampling distributions.

Bower (2003) reports that one of the misconceptions that students often have relates to the central limit theorem and is the idea that the distribution of any statistic will be approximately normal for samples of sufficiently large size. For example, the frequency histogram in figure 2.12 gives an estimate of the sampling distribution of the sample median in samples of size $n = 75$, taken with replacement from the population shown in figure 2.8. The estimate is based on 1,000 samples. The histogram shows clearly that the sampling distribution of the sample median is not symmetric and so cannot be normal.

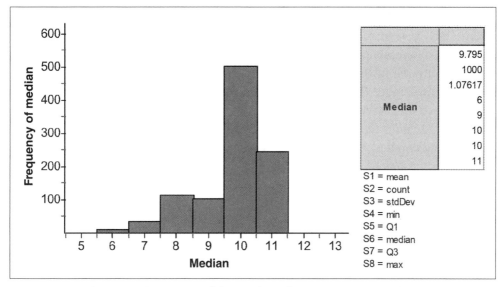

Fig. 2.12. A frequency histogram of the median of 1,000 random samples drawn with replacement, sample size $n = 75$, from the population shown in figure 2.8

Garfield and Ben-Zvi (2007) suggest that students' understanding of distributions may be increased if teachers shift students' focus from the mechanical construction of graphs to "making sense of the data, for the purpose of detecting and discovering patterns, for confirming or generating hypotheses, for noticing the unexpected, and for unlocking the stories in the data" (p. 19). In addition, Shaughnessy (2007) suggests that "students may benefit from conducting physical, hands-on sampling experiments in addition to performing these sampling experiments in a computer environment" (p. 977).

Garfield and Ben-Zvi (2007, p. 28) also report that students seem to learn better when teachers structure activities to help them evaluate the difference between their own beliefs about chance events and actual empirical results. If teachers first ask students to make guesses or predictions about data and random events, the students are more likely to care about and process the actual results. When experimental evidence explicitly contradicts these predictions, teachers should help students evaluate the differences. In fact, unless teachers force students to record and then compare their predictions with actual results, students tend to see their data as evidence that confirms their misconceptions about probability.

Simulating Random Samples

Saldanha (2004, p. 54) describes an activity that offers an example of a task structured to help students confront and correct misconceptions. Students were presented with a bag containing an unknown number of red and white candies. They were told to mix up the candies and, without looking, draw a sample of five from the bag. They were to record the number of red candies in their sample and to make a conjecture about what this sample told them about the population of candies in the bag. After putting the candies back into the bag, the students were to repeat this process nine more times. Students were then asked to interpret their results to say what these suggested about the bag's contents. Figure 2.13 presents a frequency histogram that could represent a student's first draw in the simulation—three red and two white candies. Reflect 2.6 offers an opportunity to consider your students' likely interpretations of this draw.

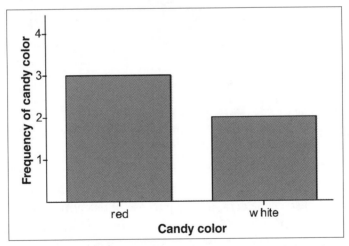

Fig. 2.13. A frequency histogram showing the numbers of red and white candies in a sample of five candies randomly drawn from an unknown population of red and white candies

Reflect 2.6

What conjectures do you think your students would make about the distribution of red and white candies in the bag if their random sample resulted in three red and two white candies?

On the basis of this one sample, students might conjecture that—

- the bag has more red candies than white candies;

- the numbers of red and white candies in the bag are about the same;

- three-fifths of the candies in the bag are red; or

- no real conjectures can be based on one small sample.

Students could then repeat the process nine more times, possibly generating the data in the table shown in figure 2.14.

Trial	Number of red candies	Number of white candies
1	3	2
2	1	4
3	2	3
4	1	4
5	0	5
6	1	4
7	2	3
8	1	4
9	1	4
10	2	3

Fig. 2.14. A table showing results for a simulation of the numbers of red and white candies in a sample of $n = 5$ candies taken from an unknown population of red and white candies, repeated for 10 random samples

From inspection of the table, students should realize that different random samples can lead to different results and that they must consider the variability inherent in the sampling process when making any inferences about the true proportions of red and white candies in the bag. Hands-on activities such as this provide students with an experiential basis for understanding technology-based simulation methods when they subsequently encounter them. As a result, when they later use technology to draw random samples, they have a better conceptual understanding of what the technology is actually doing.

After students have experienced an activity like the one described by Saldanha (2004) involving the red and white candies in a bag, they are ready for a valuable classroom discussion of an important question: "How might the information gleaned from the random samples be improved?" A common response from students to this question at this point is that the number of random samples could be increased. Because of their prior, hands-on experiences, students now know that it is difficult to make valid inferences on the basis of one small sample.

In understanding the behavior of a sample statistic, estimating the sampling distribution by taking a large number of samples is more desirable, but generating them by hand can be quite tedious. It is more efficient to use technology. For this problem, technology can quickly generate 500 or 5,000 random samples of size $n = 5$, and a histogram of the estimated sampling distribution can be constructed as in figures 2.15 and 2.16, respectively.

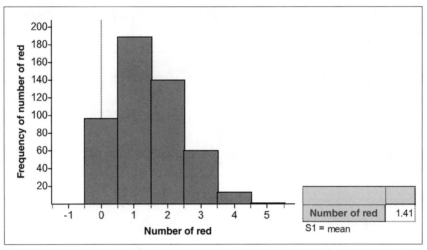

Fig. 2.15. A frequency histogram for the number of red candies in 500 random samples, sample size $n = 5$, drawn from an unknown population of red and white candies

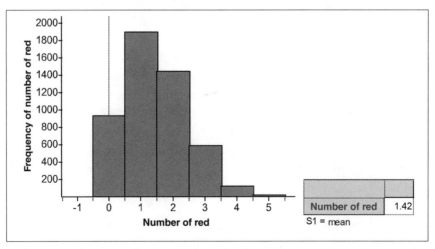

Fig. 2.16. A frequency histogram for the number of red candies in 5,000 random samples, sample size $n = 5$, drawn from an unknown population of red and white candies

It may now be clearer to students that a more accurate conjecture for the proportion of red candies in the bag would be around

$$\frac{1.4}{5} \approx 28\%.$$

(The theoretical proportion of red candies in the bag is

$$\frac{2}{7} \approx 28.6\%,$$

and the average number of red candies in each random sample of size 5 should be

$$\frac{2}{7} \cdot 5 \approx 1.43).$$

Classifying Students' Statistical Understanding

Watson and Moritz (2000) developed a three-tiered model for evaluating students' understanding of statistics, with tier 3 representing the objective for students when they leave school. Students' thinking exhibits increasing sophistication at each successive level:

- Tier 1: A basic understanding of statistical terminology

- Tier 2: An understanding of statistical language and concepts when they are embedded in the context of wider social discussion

- Tier 3: A questioning attitude ... when applying more sophisticated concepts to contradict claims made without proper statistical foundation (pp. 48–49)

With respect to the candy simulation, examples of student responses at these different levels might take the following form:

- With tier 1 skills, a student knows basic statistical terminology, and if asked a question about the definition of a random sample drawn from the bag containing red and white candies, the student might respond that it is a representative subset of the population of white and red candies in the bag.

- With tier 2 skills, a student who observes that the average number of red candies is 1.41 when 5 candies are drawn from the bag (see fig. 2.15) understands that this result does not mean that every random sample of size 5 will have 1.41 red candies but reflects what one can expect as an average.

- With tier 3 skills, a student has developed a healthy degree of skepticism. As a result, on hearing that a particular random sample has five red candies, the student might ask about whether the sample size was really $n = 5$ or how the sample was selected.

Sampling Distributions versus Population Distributions

Sampling distributions are one of the most difficult concepts for students to master, and students often have trouble differentiating among a *population distribution,* the *distribution of a sample*, and the *sampling distribution of a statistic* (Harradine, Batanero, and Rossman 2011). How you might help your students differentiate among these concepts is the focus of Reflect 2.9.

Reflect 2.9

How would you explain to your students the differences among the *population distribution*, the *distribution of a sample*, and the *sampling distribution of a statistic*?

One reason that students find sampling distributions to be much more challenging to understand than population distributions is that different units of analysis are associated with the two distributions. Even though a mean may be the statistic of interest in both distributions, the unit of analysis in the case of the population distribution is an individual object, whereas in the case of a sampling distribution, the unit of analysis is a single random sample. The means of the two distributions are interpreted differently.

For example, figure 2.17 represents 5,000 values that compose a population with an approximately exponential distribution. All 5,000 values of the population are used to compute the population mean and the standard deviation, which are 20.6 and 15.3, respectively. Each dot in the graph represents a member of the population, and the unit of analysis is an individual object.

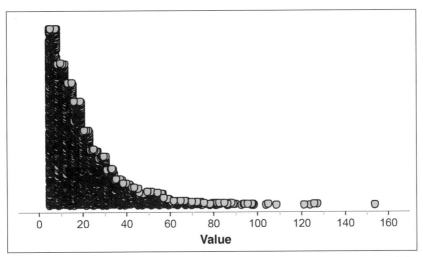

Fig. 2.17. A dot plot of a population of numbers with an approximately exponential distribution

Figure 2.18 shows a dot plot of a random sample of size 30 drawn from the population in figure 2.17. Now only 30 values of the population are available for use in computing the sample mean and standard deviation of this random sample, and these values are 19.0 and 12.3, respectively. In this case also, the unit of analysis is an individual value, and each dot in the graph represents a value that is in the random sample and also in the population.

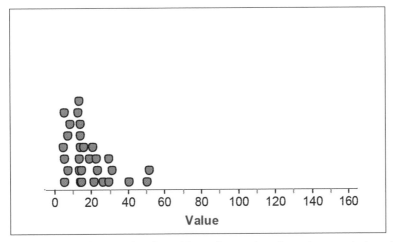

Fig. 2.18. A dot plot of a sample of $n = 30$ numbers taken from the population shown in figure 2.17

Other random samples taken from this same population can be expected to have different sample means and standard deviations. Figure 2.19 shows another sample of size 30 drawn from the population in figure 2.17. The mean of this sample is 27.7, and its standard deviation is 18.9, with each of these values computed from the 30 values in the sample. As before, the unit of analysis is an individual value, and each dot in the graph represents a value in the random sample that is also a value in the population.

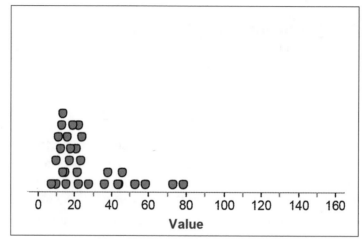

Fig. 2.19. A dot plot of a second sample of *n* = 30 numbers taken from the population shown in figure 2.17

In contrast to such cases, figure 2.20 represents the means of 50 samples, each of size 30, taken from the population in figure 2.17. Fifty sample means are available for use in computing the mean and standard deviation of this distribution, and these values are 21.3 and 2.6, respectively. But in the new situation, the unit of analysis has changed. The dots in the graph no longer represent individual values from the original population. Instead, in the graph in figure 2.20, each dot represents the mean of one of the 50 samples of size 30 drawn from the population. This distinction is very important for students to understand and is at the heart of the differences among a *population distribution*, the *distribution of a sample*, and the *sampling distribution of a statistic*.

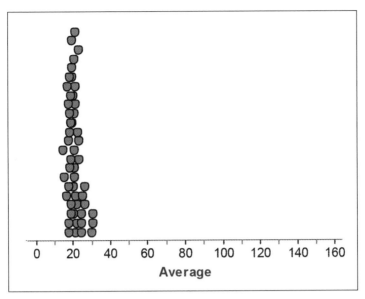

Fig. 2.20. A dot plot of 50 sample means, each based on a sample of size *n* = 30, taken from the population in figure 2.17

According to Harradine, Batanero, and Rossman (2011), students often have dif-ficulty understanding this distinction, but teachers can assist them by engaging them in activities in which they repeatedly draw and examine random samples from the same population. They must be able to connect such an activity with the question, "What would happen if I did this over and over again?" This question, according to Cobb and Moore (1997), is "the key to the logic of standard statistical inference"
(p. 818).

Furthermore, students must be able to reconcile two opposed characteristics of samples—their representativeness and their variability—when trying to understand the reason for constructing a confidence interval or conducting a hypothesis test using the information in a single random sample. Harradine, Batanero, and Rossman (2011) add that the reason for using a single sample is to judge how "rare" that sample is when compared with the numerous other samples that could have been drawn. Students who do not have a conceptual understanding of the sampling distribution of a statistic tend to judge how rare a statistic is by comparing it with the corresponding population parameter rather than with a clustering of the statis-tic's values in an appropriate sampling distribution (fig. 2.17 versus fig. 2.20).

Conclusion

The concept of variability and the interconnections between variability and sampling are complex. As this chapter has discussed, students often confuse the variability of data with the "accuracy" of individual data values, and one of the most common misconceptions that students have is the notion that a sample of any size exhibits the same characteristics as the population from which it is drawn. Working with inventing activities, predicting results and comparing the predictions later with actual results, conducting hands-on experiments, and using technology that simulates drawing multiple random samples all encourage students to examine and correct misconceptions as they quantify and interpret the variability in a data set. This work also supports students' emerging understanding of relationships among the distribution of a random sample, the sampling distribution of a statistic, and the population distribution.

Students often do not have sufficient experience with drawing samples and constructing sampling distributions. Because the concepts of sampling and sampling distributions are fundamental to an understanding of inferential statistics, students without proper exposure to these two concepts cannot be successful in understanding hypothesis tests or constructing confidence intervals when they must do more than follow a recipe to investigate and analyze a specific situation. Chapter 3 focuses on ways to develop students' understanding of hypothesis testing and the interrelated topic of data collection.

practice

Chapter 3
Data Collection and Hypothesis Testing

Big Idea 3
Hypothesis tests answer the question, "Do I think that this could have happened by chance?"

Essential Understanding 3*a*
A hypothesis test involves choosing between two competing hypotheses—the null hypothesis and the alternative hypothesis.

Essential Understanding 3*b*
The alternative hypothesis is determined by the statistical question of interest.

Essential Understanding 3*c*
The null hypothesis is rejected in favor of the alternative hypothesis if the sample data provide convincing evidence against the null hypothesis.

Essential Understanding 3*d*
The *p*-value measures surprise.

Essential Understanding 3*e*
Hypothesis tests do not always lead to a correct decision.

Big Idea 4
The way in which data are collected matters.

Essential Understanding 4*a*
Observational studies, including surveys, provide information about the characteristics of a population or sample, whereas controlled experiments provide information about treatment effects.

Essential Understanding 4*b*

Random assignment in an experiment permits drawing causal conclusions about treatment effects and quantifying the uncertainty associated with these conclusions.

Essential Understanding 4*c*

Random selection tends to produce samples that are representative of the population, permitting generalization from the sample to the larger population and also allowing the uncertainty in estimates to be quantified.

Essential Understanding 4*d*

Random selection and random assignment are different things, and the type and scope of conclusions that can be drawn from data depend on the role of random selection and random assignment in the study design.

In *A Scandal in Bohemia*, Sherlock Holmes, Sir Arthur Conan Doyle's great fictional master of deduction, says, "I have no data yet. It is a capital mistake to theorize before one has data. Insensibly one begins to twist facts to suit theories, instead of theories to suit facts." Applied to statistics, Holmes's words could serve as a warning to students about falling into the trap of the *confirmation bias*—collecting and interpreting data in such a way to support an already established viewpoint. In the realm of statistics, Holmes's observation underscores the importance of Big Idea 4, "The way in which data are collected matters," set out in *Developing Essential Understanding of Statistics for Teaching Mathematics* in Grades 9–12 (Peck, Gould, and Miller 2013).

Random Assignment and Random Selection: Recycling at Red Mountain High School, Revisited

As recounted in Chapter 2, Ms. Begay, principal of Red Mountain High School, advises the Ecology Club members to seek the assistance of the school's statistics teacher, Mrs. Lee, in setting up a trial experiment for their trash recycling project. The ensuing discussion with Mrs. Lee touches on ideas related to random assignment and random selection in a data experiment:

> After school, Morgan, Mackenzie, Antonio, and Janessa are in Mrs. Lee's classroom, explaining their project and ideas.
>
> "I understand that you have some questions about what rooms you should put the recycling bins in," says Mrs. Lee. "At one extreme, you could pick just two rooms and put a recycling bin in one of them. At the other extreme,

you could use all fifty classrooms and put recycling bins in twenty-five of them. I think you can see the problems with both of these strategies."

Do you think your students would know what problems Mrs. Lee is talking about? Reflect 3.1 focuses attention on how to address these issues with students.

Reflect 3.1

How would you explain to your students the problems associated with using only two rooms in the recycling experiment? With using all fifty rooms?

Although Mrs. Lee doesn't discuss these problems directly when she continues to talk with the students, she does address them indirectly:

> "There are two or three related issues that you will need to consider as you move forward with your study," says Mrs. Lee, "One, how many classrooms should be part of your study; two, which classrooms should you choose to be part of your study; and three, how will you determine which classrooms receive recycling bins and which classrooms do not?"

> "Regardless of how many rooms you use," she continues, "you need to make sure that you assign the 'recycling' classrooms randomly so that your results aren't biased. For example, if you decided to put recycling bins in classrooms that generate a lot of paper and not to put recycling bins in classrooms that don't generate a lot of paper, then I don't think you would have a very strong argument to convince Ms. Begay to go forward with your recycling project."

Thus, Mrs. Lee informally—even casually—introduces the idea of random assignment. Pause to think about the guidance that students might need to understand the importance of this concept, as directed in Reflect 3.2.

Reflect 3.2

How would you explain to your students why random assignment is important in the recycling experiment?

What is wrong with putting bins in the rooms that generate the most recyclable paper?

Mrs. Lee fleshes out the idea of random assignment naturally in her interaction with the students about their recycling project:

> After waiting for the students to process what she has said, Mrs. Lee explains, "Your data would be biased in favor of using recycling bins, and you would not be able to argue effectively that recycling would be generally beneficial in all of the classrooms in our school."
>
> "But Mrs. Lee, we are trying to show that recycling would be effective in our school," says Mackenzie. "What's wrong with putting bins in places where we know it would do the most good?"
>
> "If you put bins just in the rooms where recycling has the potential to be most effective, would that truly measure how beneficial a recycling program would be for the classrooms in the entire school?" Mrs. Lee asks. "What if instead you put recycling bins just in the language rooms where very little paper is used?"
>
> "But that would be unfair! They don't use much paper in those rooms—that would skew the results against what we want to show!" Morgan says indignantly.
>
> "That's right, Morgan," says Mrs. Lee, "and that's why both those ways of choosing the classrooms that will get the bins is wrong. You want the results of your study to be unbiased—your study needs to be designed so that the results are fair and do not favor any one point of view—not your point of view, nor someone else's that recycling is not needed at Red Mountain High School."
>
> After waiting again for the students to think about what she has said, Mrs. Lee sharpens her focus: "You also need to take into consideration that you'll have at least two different types of variability. First, the amount of trash in a given classroom probably isn't constant—it will vary from day to day. Second, the amount of trash you collect will probably depend on which rooms you put recycling bins in."
>
> "That makes sense to me," says Antonio.
>
> "What do you suggest, Mrs. Lee? How do we choose the rooms that will get the recycling bins so our results aren't biased?" asks Mackenzie.
>
> "One way to randomly determine which rooms will get recycling bins is to

use a calculator or computer to generate a set of random digits," Mrs. Lee responds. "As a simpler example, let's say that you had only twelve classrooms, and you wanted to put recycling bins in six of them. First, you could label the classrooms from 1 to 12. Then you could randomly generate a set of six numbers between 1 and 12. Here, let me show you."

Mrs. Lee turns and writes a calculator command rapidly on the board. "Entering '**randInt(1, 12, 6)**' into your calculator returns a list of six numbers between 1 and 12, for example"–again she writes quickly as she speaks– '{12, 11, 2, 7, 6, 4}.' With this output, you would put bins into the rooms you had labeled 2, 4, 6, 7, 11, and 12. The fact that each room had the same chance of being assigned a recycling bin would mean that your process of assigning bins to rooms had no bias."

Antonio asks, "OK, how many rooms do you think we should use? It wouldn't be fair to ask the custodial staff to collect and weigh the trash. I think members of the Ecology Club should be responsible for that."

After a quick mental calculation Morgan says, "I bet we could get enough help to handle twenty rooms."

"But Ms. Begay said we can get only four recycling bins until we can convince her that our project will work," interjects Mackenzie. "So, I guess we can use only eight classrooms and put recycling bins in four of them."

"Not necessarily," Mrs. Lee says. "You don't need to have the same number of classrooms in your two groups. The goal is to minimize the variability of your estimate of the mean amount of trash in each group of classrooms. If you have the resources to collect trash from twenty rooms, you would get the best result if you had ten rooms with recycling bins and ten rooms without recycling bins. However, you're going to have only four recycling bins. Still, you would be better off using four classrooms with trash bins and sixteen classrooms without bins than collecting trash from only eight rooms."

Mrs. Lee has introduced an idea that may surprise students: The sample sizes for the two groups in an experiment need not be the same size, and the larger the sample size can be for a group, the better. This may or may not be a notion that your students grasp intuitively. Pause to consider the question in Reflect 3.3.

Reflect 3.3

How might you convince your students that the number of rooms with recycling bins does not have to equal the number of rooms without recycling bins?

Clarifying and Reinforcing Intuition about Sample Size

Students' intuition may tell them that since the computation of the standard deviation depends on the sample size, a larger sample size will decrease the variability of the sample mean. If they do not have an intuitive grasp of this idea, they may benefit from a demonstration. Figure 3.1 presents a frequency histogram for the differences in two sample means for samples of size $n = 4$ based on 1,000 random samples from each population. This is an estimate of the sampling distribution of the difference in sample means for samples of size 4 from each population.

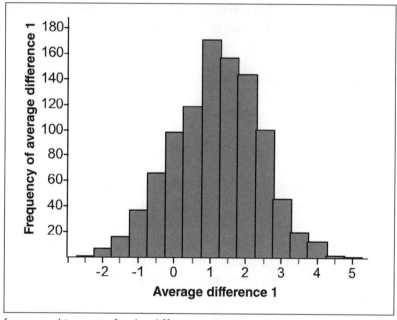

Fig. 3.1. A frequency histogram for the differences in two sample means for samples of size $n = 4$ based on 1,000 random samples from each population

Figure 3.2 also shows a frequency histogram for the difference in two sample means based on 1,000 random samples from each population. However, in this case, one sample is of size $n = 4$ and the other is of size $n = 16$. The variability in the difference in means is greater in figure 3.1 than in figure 3.2: The variance of the distribution in figure 3.1 is 1.44, whereas the variance of the distribution in figure 3.2 is 0.878. Even if students do not see much difference in the variability of the two sampling distributions, they should be able to grasp the idea that, with all other things being equal, choosing the sampling distribution that has the smaller variability will always be better than taking the one with the greater variability.

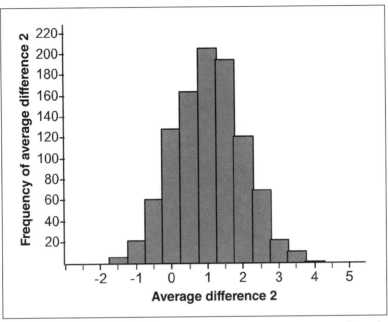

Fig. 3.2. A frequency histogram for the differences in two sample means, one based on a sample of size $n = 4$ and one of size $n = 16$ from 1,000 random samples from each population

Applying Insights: Moving the Recycling Project Forward

Once students recognize the importance of random assignment to groups and understand not only that the sample size from each group does not need to be the same but also that larger sample sizes tend to reduce variability, they can move forward in designing an experiment, as the unfolding recycling scenario demonstrates. But note the question that Antonio poses—a question that Mrs. Lee must answer first:

> "OK, Mrs. Lee, I'm convinced that it would be better to use twenty classrooms with just four with recycling bins and sixteen without, than to use just eight classrooms total," Antonio begins. "And I understand that we need to randomly assign which four get the recycling bins, but how do we decide which twenty classrooms to use in our study? Can we just choose the twenty classrooms on the first floor so it'll be less work?"

Antonio's question touches on a number of concepts. Reflect 3.4 asks you to consider the best way to respond to the question.

Reflect 3.4

How would you respond to Antonio's question?

Mrs. Lee underscores the importance of a representative random sample to obtain data on which to base valid inferences in the recycling experiment:

> "Antonio, that's a good question," says Mrs. Lee, "but it would be a mistake to choose twenty classrooms to be in the study just because they are more convenient to use. Remember, the goal of your study is to try to obtain data that will support your case about the need for recycling bins in all fifty classrooms at Red Mountain High School. To do that, the classrooms you choose should be a representative sample from among all the classrooms where you would put recycling bins at the school."

> "Well, I still think it would be, like, totally wrong to have a bin in the language rooms," grumbles Morgan.

> "That's a fair point," says Mrs. Lee, "and maybe you want to exclude the language rooms from your study entirely. But for your study you would still want to choose twenty classrooms that are representative of the remaining classrooms. This will allow you to generalize your results reasonably from the classrooms included in the study to all of the classrooms in the school."

> "How should we go about choosing the twenty classrooms?" Morgan asks.

> "I know," offers Janessa, who up to this point has been very quiet. "We can choose the classrooms randomly, by numbering them 1 to 50 and then having a calculator generate a list of twenty numbers like you showed us before!"

> "Don't forget to exclude the four language rooms," adds Morgan.

> "Right, so number the classrooms 1 to 46, and then generate a list of twenty numbers for the classrooms that will be part of the study," says Janessa.

> "Great job, kids! I think you're well on your way to implementing your study," says Mrs. Lee.

> "Oh!" she adds suddenly. "I almost forgot. There is one last thing you want

to consider as you move forward. For the classrooms that will have a recycling bin, where in the classroom are you planning to put the bin?"

"I didn't think about that," said Antonio. "Why would that matter?"

After a short silence, Morgan says, "I think I know. It's related to bias, isn't it?"

She explains, "If we put the recycle bins far away from the trash can, like in the back of the room, far from the trash can, which is near the door, then kids just won't bother going all the way to the back to do recycling. We need to make it just as easy for them to throw stuff in the recycling bin as it is for them to throw stuff in the trash can."

After more thought, she adds, "Plus, we want to make it the same in each room, so the recycling bins are not hidden back in the corner in some rooms and in other rooms are near the door."

Mrs. Lee smiles. "That sounds good, Morgan. You might also want to talk to the teachers in each of those classrooms so that they know what you are doing and agree to support your study."

"We can definitely do that," Mackenzie exclaims confidently.

"Sounds like you have a good plan," says Mrs. Lee.

Levels of Thinking about Sampling

Concepts associated with sample data are some of the most difficult statistical ideas for students to master. Hands-on experiences with drawing random samples followed by activities using technology to simulate random sampling can pave the way for students to construct a solid foundation for understanding the characteristics of sample data, especially the variability of sample distributions.

Chapter 2 referred to and briefly discussed the three-tiered classification of students' understanding of statistical concepts developed by Watson and Moritz (2000). These researchers also reported on how the classification can be applied to students' understanding of sampling. Watson and Moritz subdivided tiers 1 and 2 of their classification scheme into categories that give a finer-grained description of the characteristics of students' thinking within the tiers. Their study classified a student as a "small sampler," a "large sampler," or an "equivocal sampler," according to whether the student thought that a small number of observations (15 or fewer) were sufficient to represent a population, thought that a large number of observations (more than 20) were necessary to represent a population, or demonstrated characteristics of both small and large samplers. Watson and Moritz (2000) describe their six levels of classification as follows:

Tier 1–Understanding Terminology

Small samplers without selection (category 1)

- Students who are developing a basic concept of sample without any clear consideration of size or selection. Such students often describe samples in terms of their own naïve experience (for example, a sample of food or a small amount or subset of something).

Small samplers with primitive random selection (category 2)

- Students who are developing a basic concept of sample and may suggest what selection by a "random" process means without description of the general selection process (for example, scientists pick a sample of a few students from each school to study).

Tier 2–Understanding Terminology in Context

Small samplers with preselection of results (category 3)

- Students whose concept of sampling includes the understanding that a sample is a subset that is used to infer what the whole is like. However, they tend to want to select "average" participants instead of those with a spread of the characteristic to be studied. Although selecting only "average" participants can mislead, sampling to ensure that the spread is covered can also be incorrect if accomplished through a deliberate (as opposed to random) method of sampling.

Equivocal samplers (category 4)

- Students whose understanding of sampling includes a mixture of the characteristics of both small and large samplers and therefore cannot be classified as either.

Large samplers with random/distributed selection (category 5)

- Students who recognize that a sample is a small subset of a population but also have a notion of average in connection with the goal of a research study. They realize that people should be selected in such a way as to be representative of the population, but they do not have a proper understanding of potential bias in sampling methods and thus may select people from several different groups.

Tier 3—Critical Questioning of Claims Made without Justification

Large samplers sensitive to bias (category 6)

- Students who recognize that a sample is a small subset of a population, have a notion of average in connection with the goal of a research study, and realize that representativeness and avoiding bias are important issues in sampling. This is the goal level for students who have achieved a thorough understanding of sampling. (pp. 56–63)

Watson and Moritz determined that secondary school students (grade 9) in their study could be classified in categories 4, 5, or 6. This result suggests that many students may need experiences that allow them to develop a better understanding of the role that sample size plays in sampling methods, the ways in which samples can be biased, and the effect that biased samples can have on making statistical inferences.

The Need for Hypothesis Testing

Hypothesis testing is clearly an important statistical concept, although the Common Core State Standards for Mathematics (CCSSM) refer to it only tangentially, in the content standards set out for high school (National Governors Association Center for Best Practices and Council of Chief State School Officers [NGA Center and CCSSO] 2010, p. 82; High School.Statistics.IC.4-5):

Make inferences and justify conclusions from sample surveys, experiments, and observational studies

4. Use data from a sample survey to estimate a population mean or proportion; develop a margin of error through the use of simulation models for random sampling.

5. Use data from a randomized experiment to compare two treatments; use simulations to decide if differences between parameters are significant.

Many students have misconceptions related to inferential statistics and have difficulty in understanding null and alternative hypotheses, p-values, and levels of significance, as well as in answering the question, "Do I think that this could have happened by chance?"

Some researchers in statistics education believe that exploratory data analysis must come before formal inference. As Cobb and Moore (1997) state, "Most real data contain surprises, some of which can invalidate or force modification of the inference that was planned" (p. 808). They add that novices often make the mistake of submitting data to inference procedures before carefully exploring them. Doing so can cause the inferences to be meaningless because of questions about the data or flaws in the way that they were collected. (Questions to answer include the following: "Were the data collected through observation or experimentation?" "Do the data consist of independent measurements?" "What is the distribution from which the data are collected?") A table drawn from Cobb and Moore (1997, p. 808) and shown in figure 3.3 illustrates some of the differences between exploratory data analysis and statistical inference.

Exploratory data analysis	Statistical inference
Purpose is unrestricted exploration of the data, searching for interesting patterns	Purpose is to answer specific questions, posited before the data were produced
Conclusions apply only to the individuals and circumstances for which we have data in hand	Conclusions apply to a larger group of individuals or a broader class of circumstances
Conclusions are informal, based on what we see in the data	Conclusions are formal, backed by a statement of our confidence in them

Fig. 3.3. A table showing differences between exploratory data analysis and statistical inference. From Cobb and Moore (1997, p. 808).

Peck, Gould, and Miller (2013) do not directly address the level of significance α in Big Idea 3 or the essential understandings associated with it. Nor do the content standards in CCSSM. However, the curriculum for AP Statistics does address the concept of significance level. We include some discussion of it in this chapter, since it figures indirectly in Essential Understanding 3c: "The null hypothesis is rejected in favor of the alternative hypothesis if the sample data provide convincing evidence against the null hypothesis." The significance level α can usually be thought of as the standard against which evidence supporting the null hypothesis is judged.

Sotos and colleagues (2007, 2009) report that students' three most common misconceptions about inferential statistics relate to understanding the concept of a hypothesis test, interpreting a p-value, and interpreting the significance level α. More specifically, they say that students often incorrectly believe the following:

- A hypothesis can refer both to a population and a sample.

- The result of a hypothesis test proves the null hypothesis to be true or the alternative hypothesis to be false with 100% certainty.

- Results that are statistically significant at level α prove the falseness of the null hypothesis.

- α is the probability of making a mistake.

- $1 - \alpha$ is the probability that the null hypothesis is true.

- The p-value is the probability of making an error when rejecting the null hypothesis.

Reflect 3.5 provides an opportunity to examine these misconceptions more closely.

Reflect 3.5

What is problematic about each of the statements in the list above that causes it to be called a "misconception"?

How can you help your students overcome these misconceptions?

Some of these misconceptions may occur because students mistakenly believe that the incorrect conditional statement "If you reject the null hypothesis, the p-value is the probability that you are making the wrong decision" is equivalent to the correct conditional statement "If the null hypothesis is true, the p-value is the probability of obtaining a value of the test statistic at least as extreme as the value that was observed."

Sotos and colleagues (2007) suggest that one reason for these misconceptions is that "performing these tests requires students to understand and be able [simultaneous-ly] to relate many abstract concepts such as the concept of a sampling distribution, the significance level, null and alternative hypotheses, the p-value, and so on" (p. 108). Philosophical differences in the development of methods of hypothesis testing may also be causing some of the confusion. One viewpoint, credited to R. A. Fisher (Sotos et al. 2009), of how experimenters should evaluate hypothesis tests is that they should report only the *p*-value of the test. This view allows the experimenter to decide whether he or she is willing to reject the null hypothesis on the basis of that *p*-value. Under a second, contrasting, philosophy, credited to Neyman and Pearson (Sotos et al. 2009), the experimenter sets the level of significance α before carrying out the hypothesis test. This latter notion is closely connected with the concepts underlying type I and type II errors. These ideas are further confounded by the fact that they are not typically conveyed in the form of conditional statements. For example, the statement, "α is the probability of making a type I error," does not emphasize the reference to a conditional probability, whereas the more complete, correct statement, "If the null hypothesis is true, then α is the probability of rejecting it," does.

Harradine, Batanero, and Rossman (2011) add that many students are able to carry out the mechanics—that is, use formulas and calculate various quantities—associated with inferential statistics, but their strongly held misconceptions prevent them from being able to interpret inferential results properly. These researchers found that students were often unable to say whether the results of an experiment supported the null hypothesis, did not support the null hypothesis, or neither supported nor did not support the null hypothesis. Furthermore, they found that students had greater difficulty interpreting *p*-values appropriately than they did confidence intervals, and they concluded that "there are benefits of teaching inference via confidence intervals rather than hypothesis tests" (p. 241). Reflect 3.6 probes your thinking about this conclusion.

Reflect 3.6

Why might students have less success understanding *p*-values than confidence intervals?

Students must understand that they cannot equate the value of a statistic calculated from a single random sample with the value of the population parameter, because of the variability associated with taking random samples. For example, figure 3.4 shows the graph of and associated statistics for a random sample of size 10 taken from the population in figure 2.8. The mean of this random sample is 12.5. However, another random sample of size 10 could be taken from the same population and give the results shown in figure 3.5. The mean of this sample is only 6.6. Thus, to claim that the population mean is exactly equal to either of these sample means—or any other given sample mean—would be foolhardy.

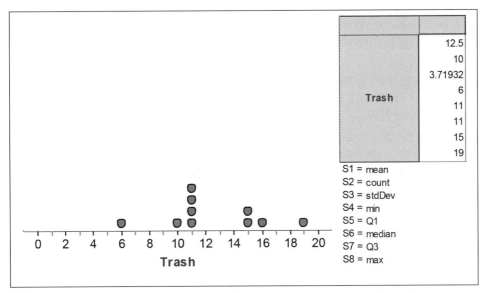

Fig. 3.4. A dot plot and descriptive statistics of a random sample of *n* = 10 numbers, taken from the data in figure 2.8

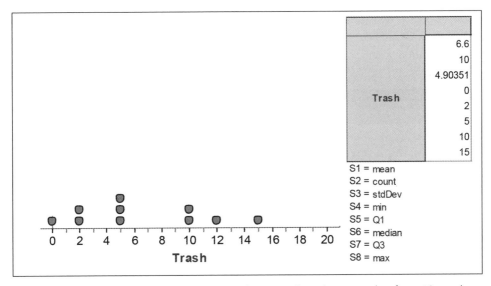

Fig. 3.5. A dot plot and descriptive statistics of a second random sample of *n* = 10 numbers, taken from the data in figure 2.8

This example also can be used to illustrate Essential Understanding 3*e*: "Hypothesis tests do not always lead to a correct decision." The results of a simulation-based hypothesis test to compare the means of the population from which the samples in figures 3.4 and 3.5 were drawn, based on resampling from the population in figure 2.8, would suggest that these samples were collected from populations with different means (p-value = .004). However, this conclusion is obviously wrong, since both samples were drawn from the same population, shown in figure 2.8.

Because the value of a statistic calculated from a single sample cannot definitively determine the value of a population parameter, students have two options from which to choose. If they want to test the evidence provided by a random sample against some claim about the population, they can use a hypothesis test. If they want to estimate a population parameter, they can construct a confidence interval. As emphasized earlier, teachers must give students opportunities to experience the variability of sample distributions before they can expect them to understand this critically important concept. Furthermore, several researchers in statistics education (for instance, Chance, delMas, and Garfield [2004]) have reported that although computer applets can help students observe patterns in sample data—seeing, for example, that variation decreases as sample size increases—the technology does not always help students understand why the patterns are occurring or help them distinguish between the distribution of a sample and the sampling distribution of means.

Rossman and Chance (2000) suggest that tests of significance should be reported in terms of p-values rather than rejection regions. Reflect 3.7 poses a question to explore this idea.

Reflect 3.7

Why might using a p-value be better than using an α-value or rejection regions in hypothesis testing?

To understand why using p-values might be better than using α-values or rejection regions, consider the following situation:

> A clinical trial is conducted to determine whether an experimental drug is effective in reducing the amount of time that an athlete needs to heal after a muscle strain to less than three weeks. One hundred four athletes with similarly severe muscle strains are divided into two groups. Sixty-eight athletes are given the experimental drug, and thirty-six athletes are given a placebo. Twenty-four athletes who received the experimental drug healed in less than three weeks, and six athletes who received the placebo healed in less than three weeks.

Would you conclude that the experimental drug is effective? Would your conclusion be the same or different if thirty-three athletes who received the experimental drug healed in less than three weeks and six athletes who received the placebo healed in less than three weeks?

In both cases, you would reject the null hypothesis that the experimental drug is ineffective at the $\alpha = .05$ level. However, the p-value for the first scenario is .046, whereas it is .0014 for the second. So, when using an $\alpha = .05$ level of significance, you would make no distinction between these two situations: In both cases, you would report that your null hypothesis would be rejected in favor of the alternative hypothesis. However, if you were using p-values, you would clearly be able to make a stronger argument for rejecting the null hypothesis in the case of the second sample than in the case of the first. Results reported in terms of the p-values are much more informative than a simple statement of rejection.

Students' attempts to distinguish between statistical and practical significance point to another common area of misconception. Sotos and colleagues (2007) elaborate:

> One of the major misconceptions mentioned in the literature, that students might encounter when evaluating hypotheses tests, is that of understanding the difference between statistical and practical significance. ... Profound knowledge about the contextualization of the test and the design of the experiment (sample sizes, etc.) as well as the encountered effect sizes, are needed in order to know when a statistically significant result is also practically significant. In turn, a practically significant result might turn out not to be statistically significant. This is the reason why... [the suggestion has been made] to drop the word *significant* from data analysis vocabulary and use it only in its everyday sense to describe something actually noteworthy or important. (pp. 106–7)

Students need to understand that statistical significance depends on sample size. For example, in comparing proportions from two samples where $\hat{p}_1 = {}^8/_{30} = .267$ and $\hat{p}_2 = {}^6/_{30} = .200$, the 6.7 percentage points between the two sample proportions is not a statistically significant difference (p-value = .54.) However, if the proportions from the two samples were $\hat{p}_1 = {}^{80}/_{300} = .267$ and $\hat{p}_2 = {}^{60}/_{300} = .200$, the difference, although still 6.7 percentage points, may or may not be deemed statistically significant (p-value = .054). But the proportions $\hat{p}_1 = {}^{120}/_{450} = .267$ and $\hat{p}_2 = {}^{90}/_{450} = .200$ would probably be deemed statistically different (p-value = .018). In each case, the difference is 6.7 percentage points, but as the sample size increases, the p-value decreases, and at some point, for a sufficiently large sample, the difference in proportions would be judged statistically significant.

Whether or not a difference of 6.7 percentage points is judged to be practically significant is context dependent: An increase in the average exam score of 6.7 percentage points might be of little practical benefit to a student if the change represents an increase from 45.0% to 51.7%. But a decrease in infant mortality of 6.7%—say, from 26.7% to 20%—might be viewed as a major accomplishment. Practical significance is a subjective judgment, not a statistical issue.

The Usefulness of Confidence Intervals

Some researchers in statistics education—Fiddler and Cumming (2005), for example—have reported benefits to using confidence intervals instead of hypothesis testing to teach statistical inference. Using confidence intervals can be a valid way of introducing ideas of statistical inference about a population parameter, even though inferences based on confidence intervals and those based on formal hypothesis-testing methods do not always lead to the same conclusions (as in proportion problems), because of some technical details. Students who were asked to interpret the results of a study were more successful if the results were presented to them as confidence intervals than if they encountered them as p-values. Further, students who were initially presented with the results as confidence intervals were then more successful in interpreting the results when presented with p-values than were students who were presented only with p-values. However, research has found no benefit to having students see p-values before seeing confidence intervals.

Unfortunately, students also harbor misconceptions associated with confidence intervals. One of the most prevalent of these is that an x% confidence interval means that the probability is $x/100$ that the confidence interval contains the population parameter. Reflect 3.8 asks you to take time to examine this misconception more closely.

Reflect 3.8

How would your students respond to the statement, "A 95% confidence interval means that the probability is .95 that the confidence interval contains the population parameter"?

How would your students interpret a 100% or a 0% confidence interval?

Applying a probability statement to the population parameter implies that the parameter can vary. But it cannot vary, since it is a constant, even if it is unknown. It is the sample statistic (used to estimate the parameter) whose value varies from random sample to random sample.

Figure 3.6 provides an example that serves to illustrate this idea. The figure shows ten 95% confidence intervals, constructed from ten random samples, each of size $n = 50$, drawn from a population of red and white candies in which the actual proportion of red candies is $p = .75$. The first of these confidence intervals estimates p to be between .64 and .88, the second confidence interval estimates p to be between .68 and .92, and the seventh one estimates p to be between .76 and 1. Nine of the ten confidence intervals shown in the figure capture the true population proportion of .75.

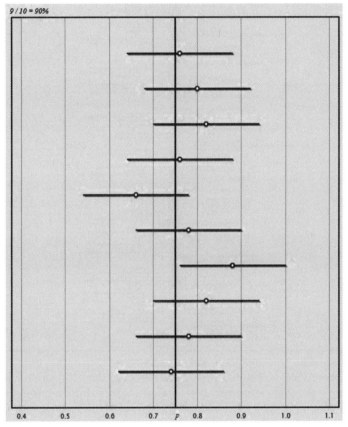

Fig. 3.6. Ten 95% confidence intervals, constructed from ten random samples, each with sample size $n = 50$, for a proportion $p = .75$ of red candies from a population of red and white candies. Constructed by using StatKey; http://lock5stat.com/statkey/index.html.

Figure 3.7 shows 100 confidence intervals constructed from 100 random samples, each of size $n = 50$, drawn from the same population of red and white candies with $p = .75$ representing the proportion of red candies. Here, 97 of the 100 intervals capture the true population proportion of .75. These two figures vividly and powerfully illustrate the meaning of a 95% confidence interval: If you could construct confidence intervals for all possible random samples of size $n = 50$, you would expect 95% of them to capture the true population proportion of red candies of .75.

This is typically a difficult concept for students to understand; they will need numerous experiences with confidence intervals to master it.

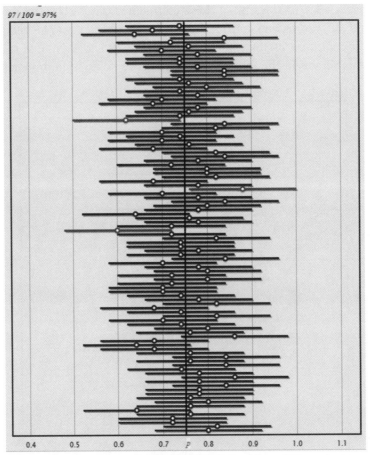

Fig. 3.7. One hundred 95% confidence intervals constructed from 100 random samples, each with sample size $n = 50$, for a proportion $p = .75$ of red candies from a population of red and white candies. Constructed by using StatKey; http://lock5stat.com/statkey/index.html.

In practice, the true population parameter is unknown. Therefore, a confidence interval based on a single random sample is used to estimate the population parameter, and the confidence placed in the estimate is based on the theoretical percentage of all possible confidence intervals that contain the actual parameter.

Students should also understand that the width of a confidence interval is directly related to the chosen level of confidence: One can increase the confidence level by increasing the width of the confidence interval. You could have a 100% confidence interval for a proportion p if you said simply that p was between 0 and 1. You would have 100% confidence in the estimate—but the estimate would not be very useful. By contrast, you could estimate p to be within a very small interval, but then you would have very little confidence in the estimate.

For example, suppose that $n = 64$ and $\hat{p} = \frac{36}{64} = .5625$. Although a 95% confidence interval for the true population proportion p would be (.441, .684), a 10% confidence interval would be (.555, .570). Reflect 3.9 focuses on another confidence interval for this example and how students might interpret it.

Reflect 3.9

In the case of the preceding example of a population of $n = 64$ and $\hat{p} = \frac{36}{64} = .5625$, how do you think your students would interpret a 0% confidence interval for the population proportion?

How would you interpret it?

Fiddler and Cumming (2005, pp. 2–3) identified several misconceptions about confidence intervals, including the following:

- A confidence interval is a descriptive statistic and has no inferential nature.

- The width of a confidence interval increases as the sample size increases.

- The width of a confidence interval is unaffected by the sample size.

- The probability is around .95 that a replication mean [that is, the sample mean from a new sample of size n] will fall within an original 95% confidence interval.

Inferential Concepts in Action: A Return to the Recycling Scenario

To see how some of these ideas might emerge and develop in a school setting, we return to the developing recycling experiment at Red Mountain High School. Morgan, Mackenzie, and Antonio are back in the science room talking with Mr. Diaz:

"Mrs. Lee has given us some good ideas about how to conduct our experiment, and we have discussed our plan with the other club members." says Morgan. "She told us that we have to be careful about how we collect the data. To make any conclusions about whether recycling bins make any difference in the amount of trash, we have to do a controlled experiment with random assignments."

Mr. Diaz nods, and Morgan continues: "Luckily, Ms. Begay told us that she found four more recycling bins, so we have a total of eight. We're going to collect and measure the trash from twenty classrooms. We will randomly assign eight rooms and put recycling bins in them. We will collect trash from twelve other randomly assigned rooms that won't have recycling bins. In the classrooms with bins, we'll make sure that the teachers know about our study, and we'll put the recycle bins right next to the trash bins, so students know they are there."

Mackenzie interjects, "And since the amount of trash will vary from room to room, we're going to compare the average amount of trash collected from the rooms. We really want to get going on this. We're starting to collect trash next week."

After the experiment begins, the club members quickly find themselves immersed in data that they set to work to organize and analyze. Later that month, Morgan, Mackenzie, and Antonio are discussing their findings with the rest of the Ecology Club. Antonio shows two dot plots (see figs. 3.8 and 3.9) of examples of the data.

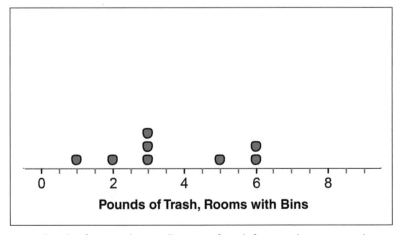

Fig. 3.8. Antonio's dot plot for one day's collection of trash from eight rooms with recycling bins

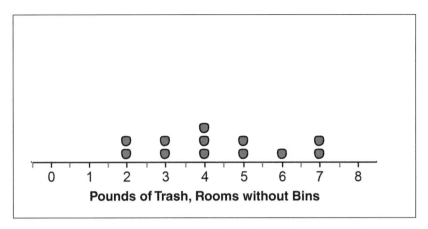

Fig. 3.9. Antonio's dot plot for one day's collection of trash from twelve rooms without recycling bins

Pointing to his dot plots, Antonio says, "On the first day we collected these amounts of trash from the eight rooms that had a recycling bin and the twelve rooms that did not. To deal with the variation that might occur because of the different rooms, we computed the average amount of trash in the two types of rooms. There was an average of 3.6 pounds of trash in the rooms that had recycling bins and an average of 4.3 pounds of trash in the rooms that did not."

"Great! Our experiment was a success!" shouts Mia.

Suppose that you have a student who, like Mia, is very quick to draw conclusions and declare an experiment to be a success on the basis of very little data. How to respond to such a student is the question that Reflect 3.10 asks you to pause and consider.

Reflect 3.10

How would you respond to Mia's statement?

Sometimes students are capable of handling such a situation on their own, reining in one another's impulse to draw conclusions prematurely, as the unfolding scenario of the recycling project demonstrates:

"Not necessarily," says Morgan. "That's just the result from the first day. Since the amount of trash can also vary from day to day, we have to look at the averages from all fifteen days." Morgan displays a table showing the daily average amounts of trash for the classrooms with bins and the classrooms without bins (see fig 3.10) as well as dot plots made with data from the table for the "bin" data (see figs. 3.11) and the "no bin" data (see fig. 3.12).

	Average_Bin	Average_No_Bin
1	3.6	4.3
2	4.0	3.1
3	3.8	4.8
4	3.9	4.9
5	3.4	6.1
6	2.5	4.4
7	3.6	3.8
8	4.3	4.1
9	2.6	5.7
10	3.6	3.1
11	4.4	4.1
12	3.1	4.3
13	3.3	3.7
14	3.3	3.0
15	3.1	2.4

Fig. 3.10. Morgan's table showing daily average amounts of trash collected for 15 days from classrooms with bins and classrooms without bins (in pounds)

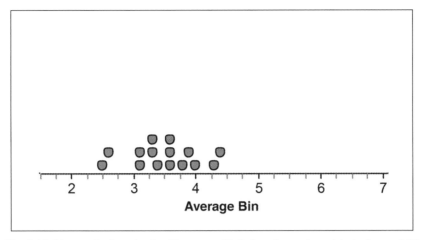

Fig. 3.11. Morgan's dot plot for "Average Bin" data from the table in figure 3.10

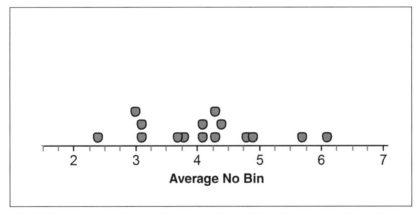

Fig. 3.12. Morgan's dot plot for "Average No Bin" data from the table in figure 3.10

Morgan continues, "The average of the daily averages for the rooms with bins is 3.50 pounds, while the average of the daily averages for the rooms without bins is 4.12 pounds."

"Great! Our experiment was a success!" shouts Mia.

Again, Mia is off in a flash to declare success. What guidance might persuade her to curb her enthusiasm? Consider the question in Reflect 3.11.

Reflect 3.11

How would you respond to Mia's statement?

Mr. Diaz draws Mia's attention to the particular sample under study, and the focus of the discussion shifts to fundamental ideas underpinning hypothesis testing and inferential statistics:

> "Not necessarily," says Mr. Diaz. "On average, the rooms without bins had more trash than the ones that had bins. However, how do you know that you didn't have that amount of trash because of your particular random sample on the fifteen days included in the study? The amount of trash varies from room to room and from day to day. It could be that if you collected the trash over a different three-week period, the amount of trash in the rooms with bins would be greater than the amount of trash in the rooms without bins. You have to decide whether you believe this result could have happened just by chance."

> "Then what can we conclude about our experiment?" asks Mia.

> Mackenzie jumps in. "This is what Morgan and I learned from Mrs. Lee. We can set up a hypothesis test. Our null hypothesis is that the average amounts of trash in the two types of rooms are the same. And we'll stick with the null hypothesis unless we have enough reason to doubt it. If we have enough doubt, then we'll reject the null hypothesis and choose our alternative hypothesis instead."

> Mackenzie pauses before continuing, getting the ideas straight: "At first, I was unsure about how to form the alternative hypothesis, so I asked Mrs. Lee if the alternative should use what she called a one-sided or a two-sided alternative. She reminded me that we needed to carefully consider how to phrase the alternative hypothesis before looking at the data. Since we believed that having recycling bins would reduce the amount of trash, our alternative hypothesis would be that the amount of trash in the rooms

without recycling bins would be greater than the amount of trash in the rooms with bins."

Mackenzie writes the null and alternative hypotheses on the board (see fig. 3.13).

$$H_0: \mu_{nonrecycling} = \mu_{recycling}$$

$$H_a: \mu_{nonrecycling} > \mu_{recycling}$$

Fig. 3.13. The null and alternative hypotheses for comparing means in the recycling experiment

"What are those funny symbols that sort of look like u's?" Mia asks.

Mackenzie explains, "That's the Greek letter mu. It's used to represent the population mean amount of trash in the two types of rooms: If all rooms didn't have recycling bins, mu of nonrecycling is what we would get for the population mean amount of trash, and if all rooms did have recycling bins, mu of recycling is what we would get for the population mean."

Morgan adds, "Using the data we collected, and assuming that our null hypothesis was true, Mrs. Lee computed a test statistic of 2.09."

She pauses and writes the number on the board. "Mrs. Lee said that the corresponding probability of observing our particular sample results or something more extreme than our results, out of all possible samples, was .024."

Morgan stops again and writes the number on the board. "She called this the p-value of our experiment."

"Actually," Morgan goes on, unfolding and holding up a sheet of paper, "Mrs. Lee made a graph—a t-curve—for me to help me understand what she said." (See fig. 3.14.)

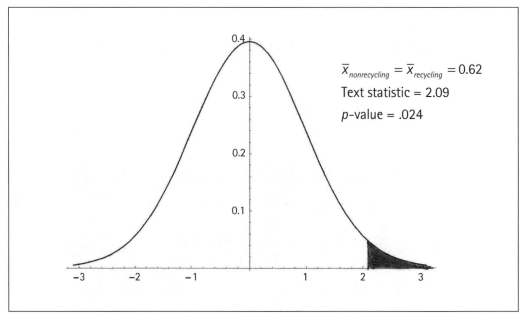

$$\overline{x}_{nonrecycling} = \overline{x}_{recycling} = 0.62$$

Text statistic = 2.09

p-value = .024

Fig. 3.14. Mrs. Lee's graph of the probability density curve for the hypothesis test for the recycling data

Morgan continues, referring to the graph: "If there is no difference in the average amount of trash collected from the rooms that do not have recycling bins and those that do, then the probability of observing a difference in sample means greater than some value x is the area under the curve to the right of x. In our experiment, the difference in sample means was 0.62 pounds. Using this value, the variability in our sample, and our sample sizes, Mrs. Lee computed a test statistic of 2.09. This means that if our null hypothesis is true, then the probability of observing our results or something more extreme is .024. You can see how far out on the right tail of the curve the test statistic of 2.09 is. If the recycling bins don't really make a difference, then our results would be very surprising. This is convincing evidence that our data are inconsistent with the null hypothesis. So we should reject it and conclude that the recycling bins do make a difference."

Footnote: Wrapping Up the Recycling Experiment

In analyzing the recycling data for the Ecology Club, Mrs. Lee used a two-sample *t*-test (estimating the variances of the two groups separately, in part, because of the unequal numbers of classrooms in the two groups, 8 versus 12), based on the fifteen daily sample means for each group (the 30 data values in the table in fig. 3.10). In doing so, she made two simplifying assumptions:

1. The amount of trash in a given classroom on one day is independent of the amount of trash on another day.

2. The amount of trash across all the classrooms on a given day is independent of the day. (That is, there are no days when the amount of trash tends to be greater in all the classrooms or other days when the amount of trash tends to be less in all the classrooms.)

A statistician analyzing these data would be likely to start with the 20 classroom measurements per day for 15 days (a total of 20 × 15 = 300 measurements) and build a statistical model to account for potential differences in classrooms with and without recycling bins, incorporating components that would allow for correlation of measurements from a single classroom across days and for correlation of measurements across all classrooms on a given day. Nevertheless, Mrs. Lee's analysis provides a reasonable assessment of the evidence supporting or refuting the null hypothesis presented in the scenario.

But after Morgan's explanation, Mia's response again is a headlong sprint toward the finish line:

> "Great! Our experiment was a success!" shouts Mia. "The null hypothesis is definitely false, and the alternative hypothesis has to be true. The recycling bins make a difference."

Reflect 3.12 poses the same question for your exploration as before.

Reflect 3.12

How would you respond to Mia's statement?

Mia needs guidance to develop her thinking and recognize the misconceptions behind her latest rush to judgment:

> "Not necessarily," says Mr. Diaz. "You have stated a common misconception. The experiment didn't conclusively prove anything. Remember what Morgan told us: If the null hypothesis is true, then the probability of seeing our results or something more extreme than our results, out of all possible samples, is .024."
>
> "Well, is it at least true that the probability is .024 of making an error if we reject the null hypothesis?" asks Mia.
>
> Mr. Diaz replies, "That's another common misconception. The p-value is not necessarily the probability of making an error in rejecting the null hypothesis. The p-value can be thought of as a measure of surprise. It is a measure of how consistent our data are with the null hypothesis. Let's look at it another way. From our data, we found an average difference of 4.12 – 3.50, or 0.62, pounds in the amount of trash collected from rooms without bins compared with the amount of trash collected from rooms with bins. Using the data you collected, Mrs. Lee also calculated a confidence interval for the difference in means. She told me that a 95% confidence interval based on the data would be (0.01, 1.23). That says we're 95% confident that the actual average difference in the amount of trash is between 0.01 and 1.2 pounds."
>
> "What does it mean to say we're 95% confident?" asks Antonio. "Does it mean that the probability is .95 that the actual average difference is between those two values?"
>
> "Not really," Mr. Diaz answers. "We could construct a 95% confidence interval every time we reran our experiment and collected a different set of data. If we could collect the data from every possible random sample of these sizes, we would find the actual difference in average weight in 95% of those confidence intervals."
>
> Antonio asks, "Well, what *can* we tell Ms. Begay?"
>
> "Hypothesis tests don't always lead to the correct decision," answers Mr. Diaz, "But you can tell her that you're fairly confident that we can decrease the amount of landfill trash from a classroom by something between 0 to 1.2 pounds per day, on average."
>
> "Great! Our experiment was a success!" shouts Mia.

If you were Mia's teacher, would you let this declaration of victory stand without challenge? This is the question that Reflect 3.13 asks you to consider.

Reflect 3.13

How would you respond to Mia's statement?

Mr. Diaz's response helps Mia to understand the limits of what she and her classmates could infer from an experiment and guides her in qualifying her sense of its "success":

> "Yes, Mia. Your experiment was a success," replies Mr. Diaz. "However, the experiment showed only that putting recycling bins into classrooms makes a statistically significant difference. Ms. Begay will still have to decide whether this difference is significant enough from a practical point of view to justify the cost of a recycling program. She may decide that the expense of buying recycling bins, collecting and bundling the recycling, and sending it to the recycling center isn't justified by the projected amount of recycling. Or she may be uncertain whether future club members will want to continue collecting the recycling, or the faculty and students will continue to recycle materials after the novelty wears off. Or she may have other concerns."

With this response, Mr. Diaz both allows Mia to feel the satisfaction of success and helps her to recognize its boundaries.

Conclusion

Collecting sample data from a population and then using sample statistics to make inferences about population parameters are foundationally important ideas in statistics. Students need extensive opportunities to collect data, both hands-on and with technology, to develop a conceptual understanding of the critical roles that sampling and hypothesis testing play in inferential statistics. As this chapter has emphasized, only through many experiences will students be able to reason through the process of developing a null and an alternative hypothesis for an experiment, evaluate their experimental results, quantify the likelihood and significance of those results, and draw inferences that are supported by statistics, while accounting for possible error. Chapter 4 extends the discussion in Chapter 3 by focusing on developing an estimator for a population parameter and evaluating the quality of the estimator.

Chapter 4
Estimators

Big Idea 5
Evaluating an estimator involves considering bias, precision, and the sampling method.

Essential Understanding 5*a*
Estimators are evaluated on the basis of their performance in repeated sampling.

Essential Understanding 5*b*
Some estimators are biased.

Essential Understanding 5c
The standard error describes the precision of an estimator.

Essential Understanding 5*d*
Confidence intervals are estimators that convey information about precision.

Essential Understanding 5*e*
The precision of estimators depends both on the way in which the sample was selected and on the size of the sample.

Essential Understanding 5*f*
If the sampling method is good, a larger sample is always more useful than a smaller sample.

Essential Understanding 5*g*
A small sample selected by using a good method can yield better results than a large sample selected by using a poor method.

Essential Understanding 5*h*
The size of the sample relative to the population size is not an important factor in determining the accuracy of estimates.

The physicist Niels Bohr has been quoted as saying, "Prediction is very difficult, especially if it's about the future." The attribution of this witty remark is open to question; others besides Bohr have been credited with the words. No matter who coined the statement, it is apt in relation to Big Idea 5 and its associated essential understandings set out in *Developing Essential Understanding of Statistics for Teaching Mathematics in Grades 9–12* (Peck, Gould, and Miller 2013). These ideas are related to what it means to have a "good" estimator, and they also describe how the sampling method and sample size can influence the usefulness of an estimator. Two of an estimator's important qualities are (1) its bias and (2) its variability.

Bias and Variability in Estimators: An Archery Analogy

Students can grasp the significance of the two attributes *bias* and *variability* in an estimator through an analogy of an archer shooting arrows at a target (see fig. 4.1). The archer's aim is *accurate* (or *unbiased*) if her arrows land in a pattern centered at the middle of the target. It is *precise* if her arrows land in close proximity to one another. Ideally, the archer would like to be both accurate and precise, with her arrows landing in a pattern that is compactly centered at the middle of the target. Such a pattern would imply not only that her repeated shots are unbiased but also that they have low variability.

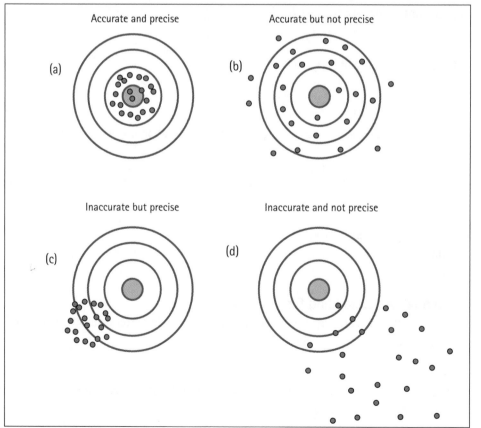

Fig. 4.1. Accuracy (bias) and precision (variability) in the landing patterns of arrows on a target

Applying the analogy to statistics, students can interpret the middle of the target as the true value of a population's parameter, and each "shot" as the corresponding statistic from a random sample. This statistic serves as an estimator of the parameter. As Essential Understanding 5a states, "Estimators are evaluated on the basis of their performance in repeated sampling," just as the archer's aim can be judged by how accurately and precisely her arrows hit the target. Her aim should be accurate—in statistical terms, unbiased—and have low variability. Before leaving the analogy of shooting arrows at a target, consider the question in reflect 4.1.

Reflect 4.1

How might students interpret each of the target patterns in figure 4.1 as an illustration of properties of an estimator of a population parameter?

The landing pattern shown on the target in figure 4.1a exhibits the desired traits of an estimator. The average value of the statistic over repeated samples is the population parameter, and the sample variability is relatively small—the value of each individual sample statistic is a reasonable estimate of the parameter. The pattern shown on the target in figure 4.1b represents an estimator for which the average value of the statistic over repeated samples is the population parameter. However, since the variability from sample to sample is great, whether the value of an individual sample statistic might be a reasonable estimate of the parameter is unknown—the values of the statistic vary widely from sample to sample.

The pattern on the target in figure 4.1c shows small sample variability, but the sample statistic is estimating something other than the population parameter, since, in this case, the average over all of the samples does not fall at the center of the target. Values of the statistic calculated from different samples consistently are too far to the left and below the target: the statistic underestimates the parameter. Similarly, in figure 4.1d, values of the statistic calculated from different samples consistently are too far to the right and below the target. The sampling procedure illustrated by the pattern on the target in figure 4.1d is not very useful. It leads to an estimate that is consistently far from the population parameter and does so with relatively large sample variability.

Using Simulation to Understand Bias in Estimators

Simulation can be used effectively to help students understand the difference between biased and unbiased estimators. Figure 4.2 presents a dot plot of a population consisting of 1,000 values that approximate an unknown theoretical distribution. The mean, median, and maximum values of this population are approximately 29.3, 30.3, and 42.0, respectively.

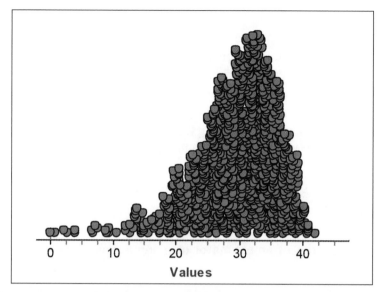

Fig. 4.2. A dot plot of 1,000 values from a population with an unknown theoretical distribution

Suppose that you take 5,000 random samples of size 5 from this population and construct the frequency histograms presented in figures 4.3, 4.4, and 4.5, which show the sample means, medians, and maximums, respectively. The mean values of these sample statistics are 29.3, 30.0, and 35.9, respectively. These frequency histograms can be viewed as estimates of the sampling distributions of the three statistics (the mean, the median and the maximum, respectively), based on samples of size $n = 5$.

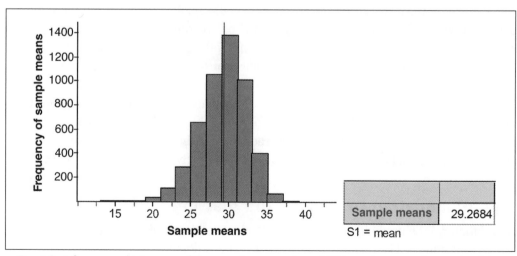

Fig. 4.3. A frequency histogram of the sample means from 5,000 random samples, each of size $n = 5$, taken from the population in figure 4.2

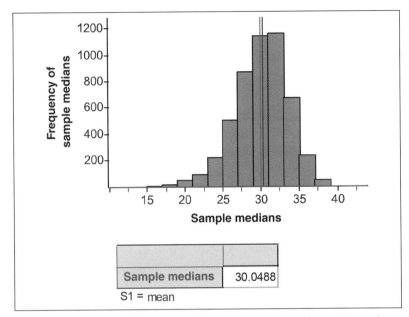

Fig. 4.4. A frequency histogram of the sample medians for 5,000 random samples, each of size $n = 5$, taken from the population in figure 4.2

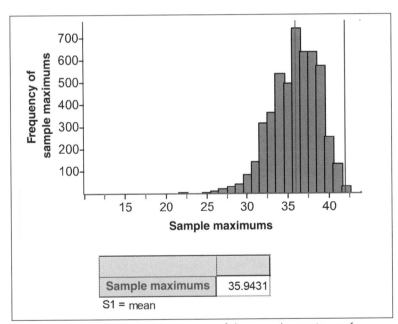

Fig. 4.5. A frequency histogram of the sample maximum for 5,000 random samples, each of size $n = 5$, taken from the population in figure 4.2

Two vertical lines superimposed on each histogram locate (1) the value of the population parameter (mean, median, or maximum) and (2) the mean of the estimated sampling distribution (frequency histogram). In figure 4.3, these two vertical lines coincide, demonstrating that the mean of the sampling distribution of sample means is the population mean. Similarly, in figure 4.4, the two vertical lines are very close to each other, demonstrating that the mean of the histogram of sample medians is very close to the median of the population. Thus, these figures permit the conclusion that the sample mean is an accurate (unbiased) estimator of the population mean, and that the sample median is an accurate (nearly unbiased) estimator of the population median. In figure 4.5, by contrast, the distance between the two lines indicates a gap between the mean of the frequency histogram (35.9) and the population maximum (42.0), demonstrating that the sampling distribution of the maximum in samples of size 5 tends to underestimate the population maximum.

Although the spread of these histograms gives some sense of the sample-to-sample variability that one can expect in the values of each of the three sample statistics, these graphs, by themselves, do not permit a judgment about whether one "can do better"—that is, increase precision (lower variability)—by using sample estimators that are different from the sample mean, median, and maximum to estimate, respectively, the population mean, median, and maximum.

Developing an Estimator of an Unknown Parameter: Raffling to Support Recycling

Understanding bias and variability in estimators is challenging for students. Figure 4.6 presents an activity set in the context of a raffle held by the Ecology Club at Red Mountain High School and designed to help students develop and strengthen their ideas about the roles that bias and variability play when using estimators of an unknown parameter. (Appendix 3 includes a student-ready activity sheet for Raffling to Support Recycling, along with notes for the teacher. The materials can be printed easily for classroom use from the appendix at nctm.org/more4u.)

The Ecology Club at Red Mountain High School is holding a raffle to raise money for a field trip to the local recycling center. The prize is a $100 gift certificate to a local movie multiplex and was donated by the theater. The club is using a roll of raffle tickets left over from last year's fundraiser, and the tickets are numbered sequentially.

Jayla and each of five friends have purchased a $1 ticket at different times over the last two weeks from the club's booth near the principal's office. They are excited about the possibility of winning the gift certificate and would like to know what their chances are. Meghan points out that if 120 tickets were sold, each of them would have a chance of 1 out of 120. So if they knew that *N* tickets were sold, they could calculate the chances of one ticket being the winner as 1/*N*, and the chances of one of the six of them being a winner as 6/*N*. Their ticket numbers are 01263, 01273, 01306, 01289, 01276, and 01399.

Kevin buys his ticket later than Jayla and her friends, and his number is 01419, but tickets are still available for sale after Kevin's purchase. When Kevin tells Jayla and her friends what his ticket number is, they realize that tickets bought later have higher numbers.

Suppose that Jayla and her friends do not know the number of the first ticket sold. They know only their own numbers and Kevin's number and that the ticket numbers are sequential. How can they use these data to estimate *N*?

Figure 4.6. Raffling to Support Recycling activity

You can guide students in simulating this problem in the following way:

1. Divide the class into groups of three or four students.

2. Give each group a paper bag containing *N* sequentially numbered "tickets" (see Appendix 3 for a template; print the tickets on cardstock if possible). For example, if *N* = 172 (the maximum number of consecutively numbered tickets possible with the template), the tickets could be numbered 01251 to 01422. Each bag should have the same number of tickets, and ideally the same starting and ending ticket numbers. The only information students should have is that the tickets in the bag are sequentially numbered. They should *not* be told the lowest ticket number.

3. Let each group randomly draw seven tickets out of the bag without replacement and record the observed numbers.

4. Ask the groups to use their seven ticket numbers to develop a method to estimate the total number of tickets in the bag.

5. Have one student from each group write the group's estimate on the board, along with a clear explanation of how the students arrived at their estimate.

6. Lead a class discussion about which estimate is best, as well as how the estimates could be tested for accuracy and variation.

The students could use any one of several different methods to make their estimates. For example, they could use the sample range, multiples of the sample standard deviation, the interquartile range, the mean, the median, or the sample maximum. Of course, students may think of other methods as well, and each of these methods will produce a sample statistic that they can use to estimate the total number of tickets. For example, assume that each bag has a total of 172 tickets and that they are numbered 01251 to 01422. One group of students might draw the numbers 01323, 01254, 01381, 01275, 01410, 01420, and 01395, as shown in figure 4.7. The table in the figure shows possible estimates for the total number of raffle tickets obtained by using some of the methods described above.

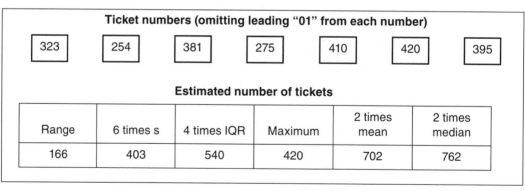

Ticket numbers (omitting leading "01" from each number)

| 323 | 254 | 381 | 275 | 410 | 420 | 395 |

Estimated number of tickets

Range	6 times s	4 times IQR	Maximum	2 times mean	2 times median
166	403	540	420	702	762

Fig. 4.7. A sample of seven ticket numbers and possible associated estimates of the total number of tickets

Note the variation in the estimates, which range from a low of 166 to a high of 762. Now the question becomes how to decide which estimate is the best one to use. A classroom discussion might generate the idea of collecting repeated samples of size 7 and examining the distribution of each particular sample statistic. Figure 4.8 shows the distributions of each of the six sample statistics shown in the table in figure 4.7, computed for 50,000 samples of seven ticket numbers randomly drawn from the paper bag.

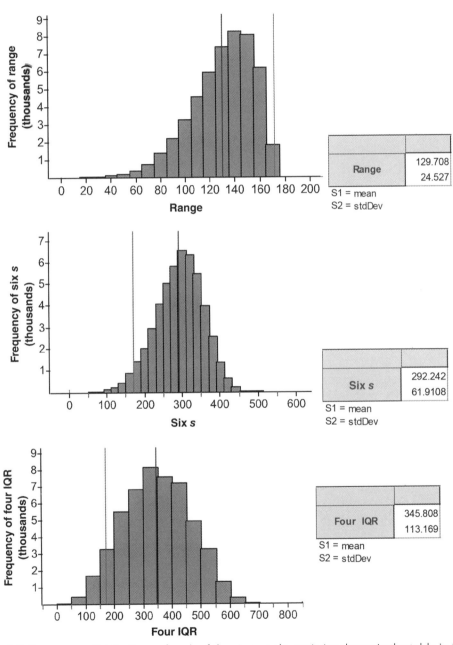

Range		129.708
		24.527

S1 = mean
S2 = stdDev

Six *s*		292.242
		61.9108

S1 = mean
S2 = stdDev

Four IQR		345.808
		113.169

S1 = mean
S2 = stdDev

Fig. 4.8. Frequency distributions of each of the six sample statistics shown in the table in figure 4.7, computed for 50,000 samples of seven randomly drawn ticket numbers

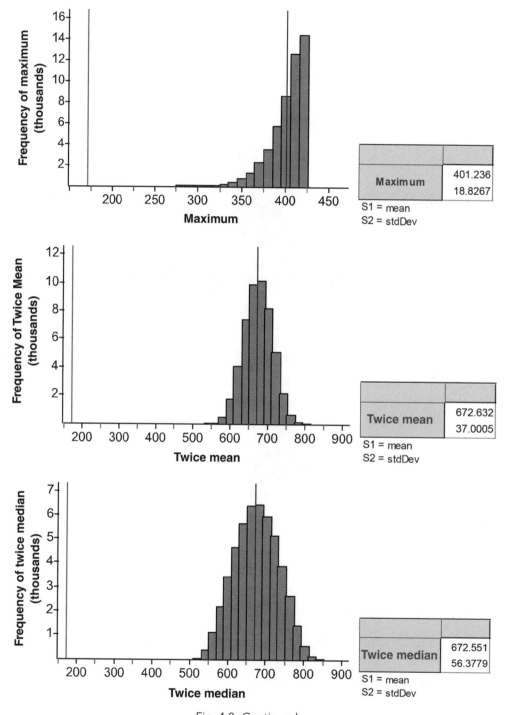

Fig. 4.8. Continued

Two vertical lines on each histogram locate (a) the total number of tickets in the bag, 172, and (b) the mean of the estimated sampling distribution of the statistic whose sampling distribution a particular histogram estimates. Reflect 4.2 offers an opportunity to explore interpretations of the frequency distributions of the six sample statistics.

Reflect 4.2

For each frequency distribution shown in figure 4.8, what does the relationship between the locations of the two vertical lines tell you about how good or poor the corresponding estimator of the total number of tickets is?

What does the spread of the histogram tell you about how good or poor the corresponding estimator is?

For an estimator, ideally the two vertical lines on the corresponding frequency histogram would be coincident, or very close to each other, indicating that on average the statistic in question is an unbiased, or nearly unbiased, estimator of the total number of tickets sold. In each of the six plots shown in figure 4.8, the vertical line at 172 and at the mean of the values in the frequency histogram are not very close to each other, indicating that none of the six estimators is especially good. The range under-estimates the total number of tickets sold, since the mean of the frequency histogram for the range is less than 172. The other five estimators all overestimate the number of tickets sold, since the mean of their respective frequency histograms is greater than 172. However, some of these estimators are better than others with respect to ac-curacy. The range on average underestimates the number of tickets sold by about 42 (172 – 129.7), whereas the estimator "6 times s" on average overestimates it by about 120 (292.2 – 172). The other estimators are even worse with respect to accuracy.

The spread of each frequency histogram provides information about how variable the estimated value is from sample to sample. This information is quantified in the standard deviation reported for each histogram. The maximum has the smallest stan-dard deviation, meaning that results are relatively consistent from sample to sample, although on average they provide an overestimate of the total number of tickets.

Influence of Sample Size on Precision and Accuracy of Estimators

Figures 4.9–4.11 can help students understand Essential Understandings 5c, 5f, and 5h. These figures show 100 sample means (n = 5, 20, and 50, respectively), each drawn from the population in figure 4.2, and they demonstrate how sample size influences the *precision* of an estimator but not its *accuracy.* The figures show visually and numerically that the variability of the sample mean decreases as the sample size increases. The measure of precision commonly used in statistics applications is the standard deviation of the sampling distribution of the means, or *standard error,* which is the quotient of the sample standard deviation (used as an estimate of the population standard deviation) and the square root of the sample size. Reflect 4.3 offers an opportunity to scrutinize the usefulness of these figures in demonstrating these concepts to students.

Reflect 4.3

How could you use figures 4.9–4.11 to convince your students of the validity of Essential Understandings 5*f* and 5*h*?

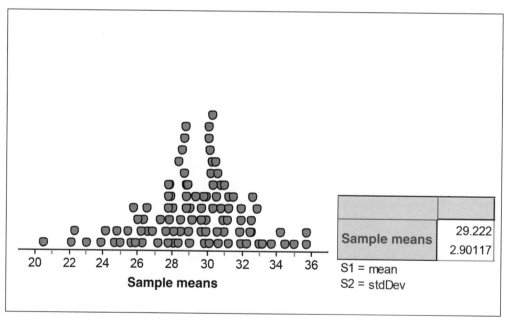

Fig. 4.9. A dot plot of the sample means for 100 random samples of size *n* = 5, taken from the population in figure 4.2

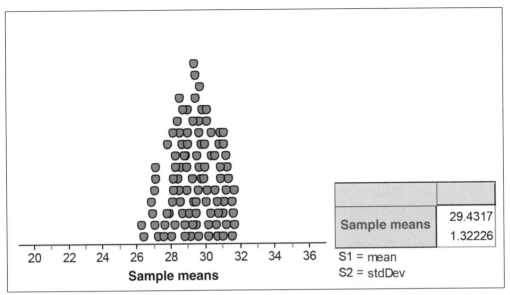

Fig. 4.10. A dot plot of the sample means for 100 random samples of size *n* = 20, taken from the population *in* figure 4.2

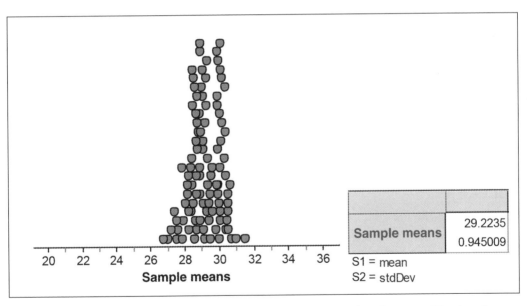

Fig. 4.11. A dot plot of the sample means for 100 random samples of size *n* = 50, taken from the population in figure 4.2

For each of the estimated sampling distributions in figures 4.9–4.11, the mean of the distribution is fairly close to the population mean of 29.3, indicating that the sample mean is a good estimator of the population mean, regardless of the sample size. Slight discrepancies in the sampling distribution means from 29.3 can be attributed to the fact that in each of these figures, the estimate of the sampling distribution is based on only 100 simulated samples, and consequently the dot plots provide a somewhat rough estimate of the corresponding true sampling distributions.

A final point concerning the use of simulation methods can help students understand the accuracy and precision of estimators: In applying statistics to analyze data, investigators get to shoot just one arrow at a target (see fig. 4.1). Their hope is that their arrow comes close to hitting the center of the target, even though they don't know where the center is located. This is to say that, in practice, investigators collect a single sample of size n (one arrow) and use the information in the sample to estimate an unknown population characteristic (the center of the target). Simulation methods are useful because they allow investigators to probe how good their shooting methods are, but when they apply statistical methods in practice, they will not be able to judge in a given instance whether their shot came close to scoring a bull's-eye or ended up somewhere off center.

Students sometimes have trouble separating the process of investigating the properties of an estimator by using simulation methods from the application of a statistical estimation method to a single sample of size n to draw conclusions about a population characteristic. For example, in trying to estimate a population mean, a student might erroneously suggest the need to take several samples of size n, and for each such sample calculate the sample mean to draw any conclusion about the value of an unknown population mean.

Conclusion

By addressing the use of a sample statistic to estimate a population parameter, Big Idea 5 and its associated essential understandings round out the set of ideas and understandings articulated for statistics by Peck, Gould, and Miller (2013). Like the foundational concepts examined in Chapters 1–3, the idea of an estimator is challenging for students to grasp. As Chapter 4 has discussed, teachers must be sensitive to students' need for carefully chosen opportunities to develop their understanding of the qualities of bias and variability in an estimator, as well as the impact of sample size and sampling method on it. Teachers must ensure that students recognize that the population parameter is generally unknown, as is how close the estimator comes to that parameter. Instruction benefits from the use of simulation that helps students see the difference between biased and unbiased estimators and hands-on activities that engage them in identifying an estimator while experiencing the effects of bias, variability, sample size, and sampling method on it. Chapter 5 takes the next critical step in expanding teachers' pedagogical knowledge for teaching statistics in grades 9–12 by setting the big ideas and essential understandings that students develop in high school in a broader context, looking at how those ideas take form in earlier years and grow in sophistication in Advanced Placement Statistics and beyond.

into practice

Chapter 5
Looking Back and Looking Ahead with Statistics

This chapter—the last in this book—looks in opposite directions to frame the discussion of students' understanding of statistics in grades 9–12. First, it examines the statistical topics addressed in grades 6–8, which lay the foundation for the understanding that students are expected to develop when they study statistics in grades 9–12. Then the chapter looks ahead and surveys the topics that students may encounter in postsecondary statistics courses. Together, these two viewpoints position the statistical concepts that students encounter and become skillful in working with in grades 9–12 in a broad context that emphasizes both their roots and their continuing growth and value.

Looking Back: Statistics in Grades 6–8

Statistical ideas have a prominent place in the content standards presented for grades 6–8 in the Common Core State Standards for Mathematics (CCSSM; National Governors Association Center for Best Practices and Council of Chief State School Offices [NGA Center and CCSSO] 2010). The standards identified below are directly related to several of the big ideas and essential understandings presented by Peck, Gould, and Miller (2013) for teaching statistics in grades 9–12. These standards for grades 6–8 lay the foundation on which students build in grades 9–12 and develop their understanding of central statistical concepts, including variability, linear regression and correlation, sampling, and inference:

For grade 6 (CCSSM 6.SP.1–5, p. 45):

Develop understanding of statistical variability.

1. Recognize a statistical question as one that anticipates variability in the data related to the question and accounts for it in the answers.

2. Understand that a set of data collected to answer a statistical question has a distribution which can be described by its center, spread, and overall shape.

3. Recognize that a measure of center for a numerical data set summarizes all of its values with a single number, while a measure of variation describes how its values vary with a single number.

Summarize and describe distributions.

4. Display numerical data in plots on a number line, including dot plots, histograms, and box plots.

5. Summarize numerical data sets in relation to their context, such as by:

   a. Reporting the number of observations.

   b. Describing the nature of the attribute under investigation, including how it was measured and its units of measurement.

   c. Giving quantitative measures of center (median and/or mean) and variability (interquartile range and/or mean absolute deviation), as well as describing any overall pattern and any striking deviations from the overall pattern with reference to the context in which the data were gathered.

   d. Relating the choice of measures of center and variability to the shape of the data distribution and the context in which the data were gathered.

For grade 7 (CCSSM 7.SP.1–4, p. 50):

Use random sampling to draw inferences about a population.

1. Understand that statistics can be used to gain information about a population by examining a sample of the population; generalizations about a population from a sample are valid only if the sample is representative of that population. Understand that random sampling tends to produce representative samples and support valid inferences.

2. Use data from a random sample to draw inferences about a population with an unknown characteristic of interest. Generate multiple samples (or simulated samples) of the same size to gauge the variation in estimates or predictions.

Draw informal comparative inferences about two populations.

3. Informally assess the degree of visual overlap of two numerical data distributions with similar variabilities, measuring the difference between the centers by expressing it as a multiple of a measure of variability.

4. Use measures of center and measure of variability for numerical data from random samples to draw informal comparative inferences about two populations.

For grade 8 (CCSSM 8.SP.1–4, p. 56):

Investigate patterns of association in bivariate data.

1. Construct and interpret scatter plots for bivariate measurement data to investigate patterns of association between two quantities. Describe patterns such as clustering, outliers, positive or negative association, linear association, and nonlinear association.

2. Know that straight lines are widely used to model relationships between two quantitative variables. For scatter plots that suggest a linear association, informally fit a straight line, and informally assess the model fit by judging the closeness of the data points to the line.

3. Use the equation of a linear model to solve problems in the context of bivariate measurement data, interpreting the slope and intercept.

4. Understand that patterns of association can also be seen in bivariate categorical data by displaying frequencies and relative frequencies in a two-way table. Construct and interpret a two-way table summarizing data on two categorical variables collected from the same subjects. Use relative frequencies calculated for rows or columns to describe possible association between the two variables.

The statistical concepts explored in grades 6–8 primarily relate to the descriptive and informal inferential statistics necessary for the more in-depth study of statistics that students undertake in grades 9–12. In the middle grades, students are expected to have an informal conception of variability and to be able to graph and describe a set of data in terms of its distribution, center, and variability. Figure 5.1 presents a graphic created by Maloney and colleagues (2014, p. 27) to show in concise form the learning trajectory for the Common Core State Standards for statistics in grades 6–8. This representation provides a useful reminder of how these topics are covered in middle school mathematics.

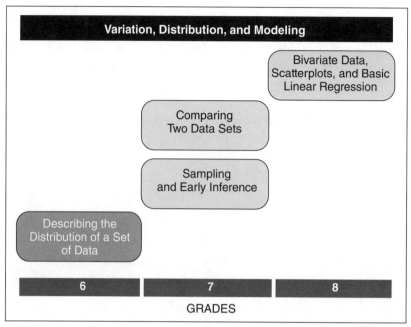

Fig. 5.1. The trajectory of learning expected by CCSSM for grades 6–8.
Adapted from Maloney and colleagues (2014, p. 27).

Working with statistics in grade 6

Encounters with statistics in grade 6 expose students to overarching ideas related to statistical questions and how measures of center and variability can be used to describe data sets. For example, students should recognize the difference between the questions, "How many text messages did I send last week?" and "How many text messages did my classmates send last week?" The second question involves variability in the data, whereas the first one does not. The second question is a statistical question, whereas the first one is not. Students in grade 6 also explore possibilities for using measures of center and variability to describe important characteristics of data sets.

Furthermore, sixth-grade students should be able to communicate relevant information about a data set graphically, numerically, and verbally. They should be able to describe the set by using its center (mean or median), spread (range, interquartile range, or mean absolute deviation), and overall pattern (distribution). See, for example, the table in figure 5.2, which presents the number of text messages that Hashem and his sixth-grade classmates sent last week, and then consider how Hashem approaches the data.

Student	Number of text messages	Student	Number of text messages	Student	Number of text messages	Student	Number of text messages
Hashem	50	Michael	25	John	32	Kayla	38
Robert	43	Kelsea	32	Salim	41	Cathleen	46
Jessica	29	Derek	45	Tyler	42	Clarissa	28
Christiana	35	Sara	50	Kiersten	39	Jerald	29
Kurren	36	Amalie	50	Zachary	44	Joshua	33
Leonardo	0	Paula	30	Carl	48	Brittany	0

Fig. 5.2. Numbers of text messages sent last week by Hashem and his classmates

Hashem plots these data in a histogram and a dot plot, as shown in figures 5.3 and 5.4, respectively. He notices that the numbers of text messages are generally spread out evenly between 25 and 50. However, he also sees that a couple of data points are different from the rest: Two of his classmates sent zero text messages last week.

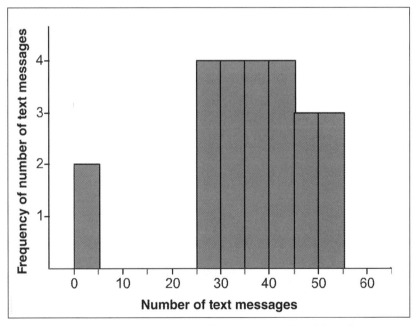

Fig. 5.3. A frequency histogram for the data in the table in figure 5.2

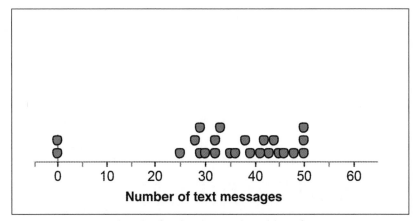

Fig. 5.4. A dot plot for the data in the table in figure 5.2

Hashem then gives a description of the data set's important features by computing and interpreting several measures of central tendency and variability for the data. Figure 5.5 shows summary measures that Hashem either computes or is given for these data.

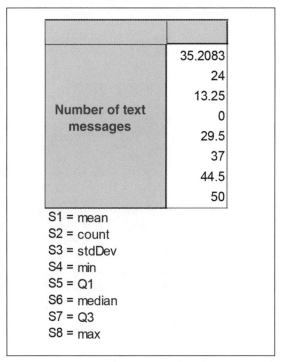

Number of text messages	35.2083
	24
	13.25
	0
	29.5
	37
	44.5
	50

S1 = mean
S2 = count
S3 = stdDev
S4 = min
S5 = Q1
S6 = median
S7 = Q3
S8 = max

Fig. 5.5. Descriptive statistics for the data from the table in figure 5.2

On the basis of these measures, Hashem is able to conclude the following:

- Last week, 50% of his classmates sent at least 37 text messages, and 50% sent 37 or fewer text messages.

- The mean number of text messages sent is 35.2, a value that Hashem can consider most representative of the number of text messages sent by his classmates.

- The interquartile range is 44.5 – 29.5, or 15.0, so the middle 50% of the number of text messages are within 15 of each other, and this information, along with the range of 50, gives an idea of the data's variability and distribution.

- Because of the two extreme values of 0, using the median to report the center might be better than using the mean.

Moreover, Hashem uses the summary statistics in figure 5.5 to construct a box plot for the data, as shown in figure 5.6. He also understands that it is important to interpret the outlier value of 0 in the context of the data so that he can answer questions like, "Are the values of zero data collection errors or data entry errors, or are they valid, though extreme, data values?" Hashem might conjecture that they represent two classmates who do not own cellphones.

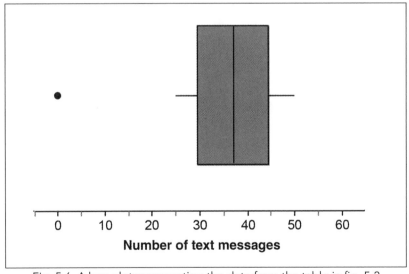

Fig. 5.6. A box plot representing the data from the table in fig. 5.2

Building statistical reasoning in grade 7

Students in grade 7 move beyond descriptive statistics and begin to look at notions related to random sampling and informal inference. For example, if Hashem is in the seventh grade instead of the sixth, and he and his classmates collect the data in the table in figure 5.2, Hashem may be interested in comparing the number of text messages sent by the girls with the number sent by the boys. Sorting the data in the table by gender results in the box plots and descriptive statistics presented in figure 5.7. Visually, as well as numerically, the variability in each of the two data sets is similar to that in the other. Hashem might note that the minimums, first and third quartiles, maximums, and standard deviations of the two data sets are about the same.

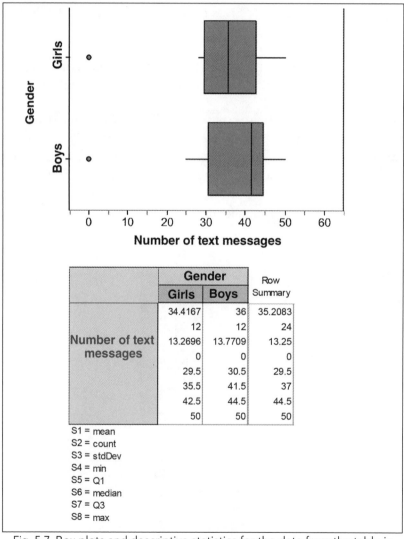

	Gender		Row
	Girls	**Boys**	Summary
	34.4167	36	35.2083
	12	12	24
Number of text messages	13.2696	13.7709	13.25
	0	0	0
	29.5	30.5	29.5
	35.5	41.5	37
	42.5	44.5	44.5
	50	50	50

S1 = mean
S2 = count
S3 = stdDev
S4 = min
S5 = Q1
S6 = median
S7 = Q3
S8 = max

Fig. 5.7. Box plots and descriptive statistics for the data from the table in figure 5.2, sorted by gender

However, the medians differ by about six messages: Last week, the boys averaged about six more text messages than the girls. Further, since the interquartile ranges of the two data sets are 13 and 14 for girls and boys, respectively, the difference in their medians is around

$$\frac{41.5 - 35.5}{13.5}, \text{ or } \frac{6}{13.5}, \text{ or } \frac{4}{9},$$

of the average of their interquartile ranges. This difference in medians is less than half the data's variability, giving good reason to believe that the difference is not significant.

Figure 5.8 presents Grade 7 Trip, a task that illustrates what students in grade 7 are expected to know and be able to apply about random sampling.

Grade 7 Trip

As president of the seventh-grade class at Stony Creek Middle School, Moriah is responsible for determining the seventh graders' preferences for their end-of-year trip. She has developed a list of five possible destinations for the trip, and she and the class's vice president, Marcus, each randomly ask 20 seventh-grade students for their preferences. They record their results, which appear in the table below. What can Moriah conclude from the data?

Results of Moriah's and Marcus's survey of destination preferences for the end-of-year trip

Moriah	Zoo	Amusement Park	Water Park	Hiking Trail	Library	Total
	2	5	10	2	1	20

Marcus	Zoo	Amusement Park	Water Park	Hiking Trail	Library	Total
	2	8	7	3	0	20

Fig. 5.8. The Grade 7 Trip task and data from two random samples of 20 seventh-grade students

Seventh-grade students should realize the importance of using a random sample—they cannot expect a sample to be representative of its population unless it is has been randomly chosen. Moriah and Marcus might get significantly different results if they merely asked some of their friends or their teammates on the cross-country team, for example. Students in grade 7 should also realize that different random samples will probably give different results, owing to the natural variability in random samples. Students should recognize that although Moriah's and Marcus's survey results differ, the difference does not mean that one of them somehow got the "wrong" results.

Furthermore, in examining Moriah's and Marcus's data sets, seventh-grade students should expect that Moriah will conclude that the amusement and water parks are probably the top choices. Students should recognize that Moriah's data show little support for going to the zoo, the hiking trail, or the library. Inspecting Marcus's data, seventh-grade students should understand that although eight students—the largest group in Marcus's sample—preferred the amusement park, that preference does not allow definitively concluding that the top preference for the entire class would be the amusement park. Indeed, seventh-grade students should note that Marcus's sample includes seven students who preferred the water park, and the difference between eight students preferring the amusement park and seven preferring the water park may simply reflect the vagaries of sampling variability.

Shifting the focus to bivariate data in grade 8

In grade 8, the statistical focus shifts to bivariate data, scatterplots, patterns of association, and constructing and interpreting lines of best fit. For example, in eighth grade, Carlos may be interested in knowing whether sitting in the front or the back of his mathematics classroom has any bearing on the grades that he receives on exams. He asks his classmates to share their grades and seating location in the room (front or back) for the previous exam and records the results, as shown in the table in figure 5.9. Consider the question in Reflect 5.1 about an eighth grader's possible interpretation of the data.

	A or B	C or lower	Total
Front of room	13	5	18
Back of room	8	10	18
Total	21	15	36

Fig. 5.9. A table showing the results of Carlos's survey on exam grade and seat location in the classroom

Reflect 5.1

What can Carlos say about whether an association exists between where he sits in the classroom and his exam grade?

Carlos might observe that approximately 72% of his classmates who sat in the front of the room made an A or a B on the exam, whereas only 44% of his classmates who sat in the back of the room made an A or a B on it. Or he might observe that approximately 62% of the students who made an A or a B on the exam sat in the front of the room, whereas 33% of the students who made a C or a lower grade on the exam sat in the front of the room. These observations might suggest to Carlos that an association exists between where a student sits in the classroom and that student's exam grade.

Given two numerical variables, students in grade 8 should be able to represent the data by using a scatterplot, make statements about how the two variables are associated, and visually insert a line of best fit for the data. For example, Bertilla's eighth-grade class is studying changes over time in the annual income that places an individual at the poverty line in the United States. In her research, she finds the data shown in the table in figure 5.10, and then she makes a scatterplot of the data, as shown in figure 5.11.

Year	Annual income ($)	Year	Annual income ($)	Year	Annual income ($)
1965	1,582	1997	8,183	2005	10,160
1970	1,954	1998	8,316	2006	10,294
1975	2,724	1999	8,501	2007	10,590
1980	4,190	2000	8,959	2008	11,201
1985	5,469	2001	9,214	2010	11,139
1990	6,652	2002	9,359	2011	11,484
1995	7,763	2003	9,573	2012	11,945
1996	7,995	2004	9,827		

Fig. 5.10. A table showing the poverty-line annual income in the United States at five-year intervals, 1965–2012. Available at http://www.infoplease.com/ipa/A0193921.html; source: U.S. Bureau of the Census, *Income, Poverty, and Health Insurance Coverage in the United States: 2008; 2010; 2012.*

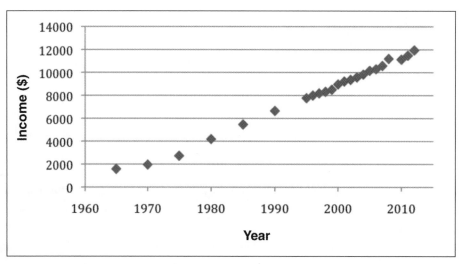

Fig. 5.11. Bertilla's scatterplot of the annual income and poverty-line data in the table shown in figure 5.10

Except for the first point, the data appear to follow a linear pattern. So Bertilla decides that the line of best fit would be the one that contains the second and last points, (1970, $1,954) and (2012, $11,945). She determines the equation of this line as approximately $y = -466,906 + 238x$ and includes its graph on her scatterplot, as shown in figure 5.12.

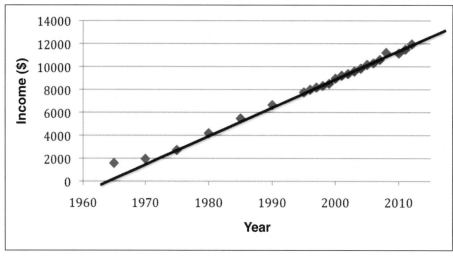

Fig. 5.12. Bertilla's scatterplot and line of best fit for data on annual income presented in the table in figure 5.10

From the equation $y = -466,906 + 238x$, Bertilla concludes that the "income poverty line" has increased at an average rate of approximately \$238 per year since 1965. She also concludes that the y-intercept of the equation has no practical implication—it would correspond to income in the year 0.

In her human biology class, Bertilla is also exploring whether any relationship exists between the age at which a child first begins talking and the child's score at a later time on a mental aptitude test. She constructs the scatterplot shown in figure 5.13 from given data.

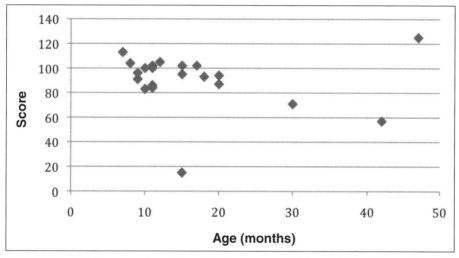

Fig. 5.13. Bertilla's scatterplot of data showing age at which a child begins to talk vs. the child's score on an aptitude test later

Bertilla first observes that the data have a generally negative association; that is, children who first begin talking at a later age tend to have lower aptitude test scores later. She also observes two data points that do not seem to follow the general trend in the rest of the data. As shown in figure 5.14, she recognizes these data points as outliers and indicates that by excluding them from the loop she draws around the other data points.

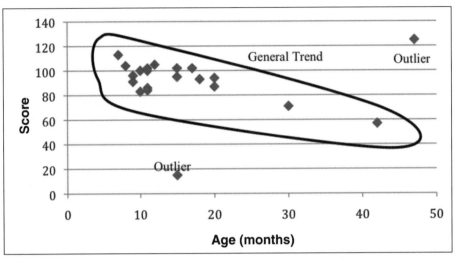

Fig. 5.14. Bertilla's representation of the general trend and her exclusion of the outliers in the data in the scatterplot in figure 5.13

As a final example of work with statistics in grade 8, consider the case of Mary, whose science class is examining the scatterplot shown in figure 5.15 of selected stars' temperatures graphed against their luminosity. Mary quickly concludes that no linear association exists between the variables, so attempting to find a line of best fit to describe the data is inappropriate.

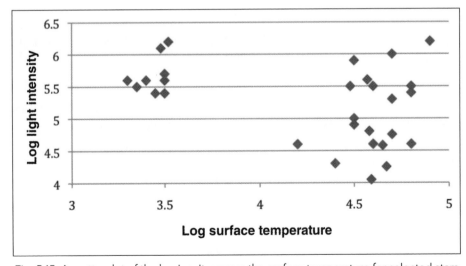

Fig. 5.15. A scatterplot of the luminosity versus the surface temperature for selected stars

However, Mary is curious about why the data seem to divide into two groups, or clusters. After further research, she notices that the cluster on the left is composed of giant stars, whereas the cluster on the right is composed of dwarf stars, as labeled in figure 5.16. These two types of stars separate naturally according to characteristics that are unobserved if one considers only the information in figure 5.15.

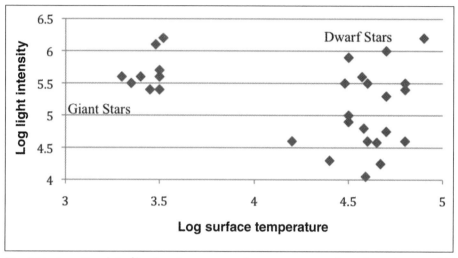

Fig. 5.16. A scatterplot of luminosity versus surface temperature for selected stars, with star types identified for the two clusters of data

Looking Ahead: Statistics beyond Grade 12

Students in college are likely to encounter further study of statistics regardless of their major. For many college students, one or more general education courses may include exposure to statistical thinking or statistical literacy. In the social and behavioral sciences, the natural sciences, and commerce and business, data collection by means of surveys, polls, observational studies, or controlled experiments is common, and students typically receive additional training in statistical thinking. In some cases these courses are discipline-specific and focus on key statistical methods and techniques within that discipline, whereas in other cases students may be required to take a general introductory statistics course focusing on broadly applicable methods of data collection and analysis.

Perhaps the common element in a statistics course at the postsecondary level is the effort to develop in students a deeper, more formal understanding of the statistical reasoning and methods than they achieved in high school. Although postsecondary statistics places substantial emphasis on developing students' informal, intuitive understanding, it also demands that students use formulas to quantify statistical relationships and conclusions. Often the focus is less on doing hand calculations with the formulas than on understanding what the formulas mean, and then using technology to perform the statistical calculations. Sometimes the technology consists of calculators or spreadsheet software, but more commonly it includes statistical computer programs such as Minitab or JMP. Furthermore, postsecondary statistics emphasizes the growth of students' ability to communicate statistical ideas and reasoning in precise written English.

Guidelines for Assessment and Instruction in Statistics Education: College Report (GAISE; Aliaga et al. 2005) recommends that an introductory course in statistics at the college level include the following:

- An emphasis on statistical literacy and statistical thinking

- The use of real data

- A focus on conceptual understanding rather than knowledge of procedures

- The use of technology for both developing understanding and data analysis

Further, GAISE recommends the use of in-class activities to foster active learning in the classroom, along with well-designed assessments to improve and evaluate student learning. In this regard, the GAISE recommendations can be seen as not substantially different from, but a natural extension of, the suggestions in this book for putting essential understanding of statistics into practice in the high school mathematics classroom.

Notably, GAISE does not suggest a list of specific statistical procedures for coverage in such a course but rather emphasizes the importance of developing students' conceptual understanding of statistics. Figure 5.17 shows common goals and emphases in postsecondary statistics courses.

Common emphases of a postsecondary course in statistics can include the following:

1. **Big Idea 4:** An understanding of the importance of random sampling and the different ways in which a random sample may be obtained (simple random, stratified, and cluster); differences between observational and experimental designs; random sampling as the basis for generalizing the results from surveys and experiments to a population; the use of random assignment in a controlled experiment to permit drawing cause-and-effect conclusions

2. **Big Ideas 1 and 2:** An understanding of variability and sampling distributions

3. **Essential Understandings 5e–5h:** An understanding of the use of statistical methods to determine sample size needs to meet objectives of a study

4. **Big Ideas 3 and 5:** An understanding of the use of specific techniques for statistical inference, including necessary assumptions and ways to check assumptions. The most common of these include using hypothesis tests and confidence intervals—

   - concerning a single population mean;

   - to compare several population means;

   - concerning a single population proportion;

   - to compare two population proportions;

   - to measure an association between two categorical variables;

   - concerning the slope of an estimated line describing the relationship between two continuous variables.

5. **Big Ideas 1–5:** An ability to interpret and communicate the results of an analysis of a set of data

Fig. 5.17. Common emphases of a postsecondary course in statistics, with connections identified to the big ideas and essential understandings for statistics presented by Peck, Gould, and Miller (2013)

The College Board AP (Advanced Placement) Statistics program consists of a course and standardized exam that is modeled on the ideal elements of a first college-level course in statistics. Students who score high enough on the exam have the potential to earn college credit for an introductory statistics course.

The AP Statistics course covers four major topic areas (adapted from College Board [2010]):

1. Exploring data (20–30%). Exploratory analysis of data makes use of graphical and numerical techniques to study patterns and departures from patterns.

2. Sampling and experimentation: Planning and conducting a study (10–15%). Data must be collected according to a well-developed plan if valid information is to be obtained.

3. Anticipating patterns: Exploring random phenomena, using probability and simulation (20–30%). Probability is the tool used for anticipating what the distribution of data should look like under a given model.

4. Statistical inference (30–40%). Statistical inference guides the selection of appropriate models.

The first AP topic, exploring data, engages students in constructing and interpreting graphical displays of data and numerical summary measures of center and spread. In addressing this topic, students work with content and develop an understanding of ideas that are similar to those outlined in the high school statistics component of CCSSM and captured in the big ideas and essential understandings presented by Peck, Gould, and Miller (2013). In addition, the list of content for this AP topic includes measures of relative standing (quartiles, percentiles, z-scores), and ways in which change in measurement scale affects numerical summary measures. Comparisons of distributions and a more formal discussion of bivariate relationships for both continuous and categorical data come under this topic as well. Students consider correlation, fit least-squares regression lines to data sets, and examine and interpret residual plots. The key difference in the coverage of this topic at the postsecondary level and in a high school setting is the addition at the postsecondary level of more formal, numerical calculations (with the explicit aid of technology) to summarize univariate and bivariate relationships.

The second AP topic, sampling and experimentation, develops the concepts captured in Big Idea 4, "The way in which data are collected matters," and its associated essential understandings. Areas of focus in this topic include the various ways of collecting data (sample survey, observational study, and experimental design), planning and conducting surveys and experiments, and generalizability of results. Increased emphasis is placed on understanding the various kinds of bias that can occur in sampling and surveys, valid alternatives to simple random sampling (cluster sampling, stratified sampling), the specific terminology involved in designed experiments, and methods for controlling for sources of variation (for example, randomized block designs and matched pairs design).

The recycling scenario presented in this book points to some of these areas. For example, the members of the Ecology Club collect data (amount of trash) for fifteen days for classrooms both with and without recycling bins. One potential source of variation in the amount of trash in a classroom might be the day of the week on which the club members collect data. For example, if the fifteen-day (three-week) period over which the students collect data includes two Fridays that are early re-lease days, they might expect less trash to be discarded on those days, regardless of whether a classroom contains a recycling bin. If the amount of trash can vary by day of the week, for whatever reason, this increases the day-to-day variability that would appear in the data.

In an AP (or college-level) classroom, students would learn that such potential sources of variability are examples of *blocking factors*. A well-thought-out experimental design would consider what blocking factors might be present and use appropriate statistical methods to *control* (that is, eliminate or remove) the variation in the re-sponse resulting from the presence of the blocking factor. In the recycling scenario, if the members of the Ecology Club want to *block on day* (in other words, take potential day-to-day differences into account), then Mrs. Lee might suggest that they use a paired *t*-test rather than the two-sample *t*-test outlined in the scenario in Chapter 3. By removing a source of variation that is a nuisance (not a variable of intrinsic inter-est) in the response, one can generally develop more precise inferences about popula-tion parameters.

The third AP topic, anticipating patterns, is closely associated with, and deepens understanding of, Big Idea 2, "Distributions describe variability." Informal methods of understanding randomness, variation, and sampling distributions based on looking at simulations of sampling distributions are extended and augmented by a more math-ematical treatment of probability, random variables, and statistical distributions. Top-ics include definitions and rules for probability and conditional probability, discrete and continuous random variables, independence of random variables, the binomial distribution, the normal distribution, the law of large numbers, and the central limit theorem, as well as an introduction to the *t*- and chi-square distributions.

The fourth AP topic, statistical inference, is the largest single component of both a typical college-level introductory statistics course and the AP course. The focus is on extending and enriching understanding of Big Idea 3, "Hypothesis tests answer the question, 'Do I think that this could have happened by chance?'" and, to a lesser extent, Big Idea 5, "Evaluating an estimator involves considering bias, precision, and the sampling method." The two main types of formal methods of statistical inference (point and confidence interval estimation, and tests of significance) are studied in a wide range of settings that might be encountered in statistical applications. This work involves students in exploring the uses of these methods identified in the bulleted list under item 4 in figure 5.17, as well as several other specific statistical methods. In addition to learning how to apply and interpret methods for statistical inference, students are engaged in a more formal discussion of the logic behind both hypothesis testing and confidence intervals, and students practice not just the mechanics of applying these methods but also techniques for correctly and effectively communicating their findings.

Two additional key components of the AP course are (1) an emphasis on the use of technology to help both in understanding and in exploring the properties of statistical methods, as well as to carry out the calculations in data analysis, and (2) an emphasis on clear communication of statistical results in plain English.

Conclusion

This chapter has demonstrated that the ideas about statistics that students encounter in high school are natural extensions and developments of ideas that they begin to learn years earlier in more elementary experiences with data. The understanding that students develop in high school is on a trajectory of learning that begins informally in the elementary grades, extends through the middle grades, grows increasingly formal in high school, and becomes yet more formal and sophisticated in AP Statistics, college courses in many diverse majors, and careers that students move into after college in many fields. By developing your understanding of the progression of these ideas, you are putting yourself in the best position to put them into practice in your classroom and your teaching, offering your students the rich experiences that they need to develop a robust, flexible, and growing understanding of statistics.

Appendix 1
The Big Ideas and Essential Understandings for Statistics

This book focuses on the big ideas and essential understandings that are identified and discussed in *Developing Essential Understanding of Statistics for Teaching Mathematics in Grades 9–12* (Peck, Gould, and Miller 2013). For the reader's convenience, the complete list of the big ideas and essential understandings in that book is reproduced below.

Big Idea 1. Data consist of structure and variability.

Essential Understanding 1*a*. Mathematical models describe structure.

Essential Understanding 1*b*. Statistical models extend mathematical models by describing variability around the structure.

Essential Understanding 1*c*. Statistical models are evaluated by how well they describe data and whether they are useful.

Big Idea 2. Distributions describe variability.

Essential Understanding 2*a*. A population distribution describes variability in the values that make up a population.

Essential Understanding 2*b*. The population distribution is often unknown but can be approximated by a sample distribution.

Essential Understanding 2*c*. The sampling distribution of a sample statistic describes how the value of the statistic varies from sample to sample.

Essential Understanding 2*d*. Simulation can be used to approximate sampling distributions.

Big Idea 3. Hypothesis tests answer the question, "Do I think that this could have happened by chance?"

Essential Understanding 3*a*. A hypothesis test involves choosing between two competing hypotheses—the null hypothesis and the alternative hypothesis.

Essential Understanding 3*b*. The alternative hypothesis is determined by the statistical question of interest.

Essential Understanding 3*c*. The null hypothesis is rejected in favor of the alternative hypothesis if the sample data provide convincing evidence against the null hypothesis.

Essential Understanding 3*d*. The *p*-value measures surprise.

Essential Understanding 3*e*. Hypothesis tests do not always lead to a correct decision.

Big Idea 4. The way in which data are collected matters.

Essential Understanding 4*a*. Observational studies, including surveys, provide information about the characteristics of a population or sample, whereas controlled experiments provide information about treatment effects.

Essential Understanding 4*b*. Random assignment in an experiment permits drawing causal conclusions about treatment effects and quantifying the uncertainty associated with these conclusions.

Essential Understanding 4*c*. Random selection tends to produce samples that are representative of the population, permitting generalization from the sample to the larger population and also allowing the uncertainty in estimates to be quantified.

Essential Understanding 4*d*. Random selection and random assignment are different things, and the type and scope of conclusions that can be drawn from data depend on the role of random selection and random assignment in the study design.

Big Idea 5. Evaluating an estimator involves considering bias, precision, and the sampling method.

Essential Understanding 5*a*. Estimators are evaluated on the basis of their performance in repeated sampling.

Essential Understanding 5*b*. Some estimators are biased.

Essential Understanding 5*c*. The standard error describes the precision of an estimator.

Essential Understanding 5*d*. Confidence intervals are estimators that convey information about precision.

Essential Understanding 5*e*. The precision of estimators depends both on the way in which the sample was selected and on the size of the sample.

Essential Understanding 5*f*. If the sampling method is good, a larger sample is always more useful than a smaller sample.

Essential Understanding 5*g*. A small sample selected by using a good method can yield better results than a large sample selected by using a poor method.

Essential Understanding 5*h*. The size of the sample relative to the population size is not an important factor in determining the accuracy of estimates.

Appendix 2
Resources for Teachers

The following list highlights a few of the many books, articles, and online resources that are helpful for teaching statistics in grades 9–12. This appendix is reproduced from *Developing Essential Understanding of Statistics for Teaching Mathematics in Grades 9–12* (Peck, Gould, and Miller 2013, pp. 114–19), where it also appears as Appendix 2, "Resources for Teachers."

The resources below have been selected to provide teachers with supplemental material to enhance their understanding of both the mechanics and the concepts of statistics, as well as to provide them with activities that they can use directly with their students in high school mathematics and statistics classes. General resources appear below, followed by specific resources targeted to requirements of the Common Core State Standards for Mathematics related to courses in algebra 1, geometry, and algebra 2. Some of the specific course resources identified below come from the sources that are listed as "general resources," whereas others are from specific books or journal articles.

General Resources

Textbooks, journal articles, and activity books

Agresti, Alan, and Christine Franklin. *Statistics: The Art and Science of Learning from Data.* 2nd ed. Upper Saddle River, N.J.: Pearson, 2009.

Beseler, Susan. "The Three-Point Shoot-Out: The Logic of Hypothesis Testing." *Mathematics Teacher* 99 (April 2006): 582–87.

Bock, David E., Paul F. Velleman, and Richard D. De Veaux. *Stats: Modeling the World.* 3rd ed. Boston: Pearson, 2010.

Bright, George W., Wallece Brewer, Kay McClain, and Edward S. Mooney. *Navigating through Data Analysis in Grades 6–8. Principles and Standards for School Mathematics* Navigations Series. Reston, Va.: National Council of Teachers of Mathematics, 2003.

Bright, George W., Dargan Frierson, Jr., James E. Tarr, and Cynthia Thomas. *Navigating through Probability in Grades 6–8. Principles and Standards for School Mathematics* Navigations Series. Reston, Va.: National Council of Teachers of Mathematics, 2003.

Burrill, Gail F., and Portia C. Elliott, eds. *Thinking and Reasoning with Data and Chance.* Sixty-eighth Yearbook of the National Council of Teachers of Mathematics (NCTM). Reston, Va.: NCTM, 2006.

Burrill, Gail, Christine A. Franklin, Landy Godbold, and Linda J. Young. *Navigating through Data Analysis in Grades 9–12. Principles and Standards for School Mathematics* Navigations Series. Reston, Va.: National Council of Teachers of Mathematics, 2003.

Cobb, George W., and David S. Moore. "Mathematics, Statistics, and Teaching." *American Mathematical Monthly* 104 (November 1997): 801–23.

Devlin, Thomas F. "Why Aren't They Called Probability Intervals?" *Mathematics Teacher* 101 (May 2008): 647–51.

Franklin, Christine, Gary Kader, Denise Mewborn, Jerry Moreno, Roxy Peck, Mike Perry, and Richard Scheaffer. *Guidelines for Assessment and Instruction in Statistics Education* (GAISE Report): *A Pre-K–12 Curriculum Framework.* Alexandria, Va.: American Statistical Association, 2007.

Gelman, Andrew, and Deborah Nolan. *Teaching Statistics: A Bag of Tricks.* New York: Oxford University Press, 2002.

Gould, Robert, and Colleen Ryan. *Introductory Statistics: Exploring the World through Data.* Boston: Pearson, 2012.

Kader, Gary, and Tim Jacobbe. *Developing Essential Understanding of Statistics for Teaching Mathematics in Grades 6–8.* Essential Understanding Series. Reston, Va.: NCTM, 2013.

Peck, Roxy, George Casella, George Cobb, Roger Hoerl, Deborah Nolan, Robert Starbuck, and Hal Stern. *Statistics: A Guide to the Unknown.* 4th ed. Florence, Ky.: Cengage Learning, in partnership with the American Statistical Association, 2006.

Peck, Roxy, Chris Olsen, and Jay L. Devore. *Introduction to Statistics and Data Analysis.* 4th ed. Boston: Cengage Learning, 2012.

Peck, Roxy, and Daren Starnes. *Making Sense of Statistical Studies.* Alexandria, Va.: American Statistical Association, 2009.

Peters, Susan A. "Engaging with the Art and Science of Statistics." *Mathematics Teacher* 103 (March 2010): 496–503.

Rossman, Allan J., Beth L. Chance, and J. Barr von Oehsen. *Workshop Statistics: Discovery with Data and the Graphing Calculator.* 3rd ed. Indianapolis: John Wiley & Sons, 2008.

Salsburg, David. *The Lady Tasting Tea: How Statistics Revolutionized Science in the Twentieth Century.* New York: W.H. Freeman, 2001.

Scheaffer, Richard L., Ann Watkins, Jeffrey Witmer, and Mrudulla Gnanadesikan. *Activity-Based Statistics*. 2nd ed., revised by Tim Erickson. Indianapolis: John Wiley & Sons, 2004.

Shaughnessy, J. Michael, Gloria Barrett, Rick Billstein, Henry A. Kranendonk, and Roxy Peck. *Navigating through Probability in Grades 9–12*. *Principles and Standards for School Mathematics* Navigations Series. Reston, Va.: National Council of Teachers of Mathematics, 2004.

Shaughnessy, J. Michael, and Beth Chance. *Statistical Questions from the Classroom*. Reston, Va.: National Council of Teachers of Mathematics, 2005.

Shaughnessy, J. Michael, Beth Chance, and Henry Kranendonk. *Focus in High School Mathematics: Reasoning and Sense Making in Statistics and Probability*. Reston, Va.: National Council of Teachers of Mathematics, 2009.

Starnes, Daren S., Daniel S. Yates, and David S. Moore. *The Practice of Statistics*. 4th ed. New York: W.H. Freeman, 2012.

Vickers, Andrew. *What Is a p-Value Anyway? 34 Stories to Help You Actually Understand Statistics*. Boston: Pearson, 2010.

Watkins, Ann E., Richard L. Scheaffer, and George W. Cobb. *Statistics in Action: Understanding a World of Data*. 2nd ed. Emeryville, Calif.: Key Curriculum Press, 2008.

Online resources

American Statistical Association. *Journal of Statistics Education*. http://amstat.org/publications/jse

American Statistical Association. Webinars. http://amstat.org/education/webinars/index.cfm

Consortium for the Advancement of Undergraduate Statistics Education. Webinars. http://www.causeweb.org/webinar

Consortium for the Advancement of Undergraduate Statistics Education. Chance News. http://www.causeweb.org/wiki/chance/index.php/Main_Page

International Statistics Literacy Project. http://www.stat.auckland.ac.nz/~iase/islp/priteach

National Council of Teachers of Mathematics. Illuminations. http://illuminations.nctm.org/WebResourceList.aspx?Ref=2&Std=4&Grd=0

North Carolina School of Science and Mathematics. Statistics Institutes. http://www.ncssm.edu/courses/math/Stat_Inst/links_to_all_stats_institutes.htm

STatistics Education Web (STEW). http://amstat.org/education/stew/index.cfm

Statistics Teacher Network. http://amstat.org/education/stn

Course-Specific Resources

Algebra 1 resources

Bohan, Jim. "Using Regression to Connect Algebra to the Real World." In *Thinking and Reasoning with Data and Chance,* Sixty-eighth Yearbook of the National Council of Teachers of Mathematics (NCTM), edited by Gail F. Burrill, pp. 195–208. Reston, Va.: NCTM, 2006.

Bright, George W., Wallece Brewer, Kay McClain, and Edward S. Mooney. *Navigating through Data Analysis in Grades 6–8. Principles and Standards for School Mathematics* Navigations Series. Reston, Va.: National Council of Teachers of Mathematics, 2003.

Burke, Maurice J., and Ted R. Hodgson. "Using Technology to Optimize and Generalize: The Least-Squares Line." *Mathematics Teacher* 101 (September 2007): 102–7.

Canada, Daniel L. "The Known Mix: A Taste of Variation." *Mathematics Teacher* 102 (November 2008): 286–91.

Erickson, Tim. "A Pretty Good Fit." *Mathematics Teacher* 102 (November 2008): 256–62.

Kader, Gary D., and Christine A. Franklin. "The Evolution of Pearson's Correlation Coefficient." *Mathematics Teacher* 102 (November 2008): 292–99.

Shaughnessy, J. Michael, and Maxine Pfannkuch. "How Faithful Is Old Faithful? Statistical Thinking: A Story of Variation and Prediction." *Mathematics Teacher* 95 (April 2002): 252–59.

Wilson, David C. "The Median-Median Line." *Mathematics Teacher* 104 (November 2010): 262–67.

Geometry resources

Albert, Jim. "Interpreting Probabilities and Teaching the Subjective Viewpoint." In *Thinking and Reasoning with Data and Chance,* Sixty-eighth Yearbook of the National Council of Teachers of Mathematics (NCTM), edited by Gail F. Burrill, pp. 417–33. Reston, Va.: NCTM, 2006.

Bright, George W., Wallece Brewer, Kay McClain, and Edward S. Mooney. *Navigating through Data Analysis in Grades 6–8. Principles and Standards for School Mathematics* Navigations Series. Reston, Va.: National Council of Teachers of Mathematics, 2003.

Danielson, Christopher, and Eric Jenson. "Probability in Practice: The Case of Turkey Bingo." *Mathematics Teacher* 102 (November 2008): 248–54.

Rubel, Laurie H. "Students' Probabilistic Thinking Revealed: The Case of Coin Tosses." In *Thinking and Reasoning with Data and Chance,* Sixty-eighth Yearbook of the National Council of Teachers of Mathematics (NCTM), edited by Gail F. Burrill, pp. 49–59. Reston, Va.: NCTM, 2006.

Scheaffer, Richard L., Ann Watkins, Jeffrey Witmer, and Mrudulla Gnanadesikan. *Activity-Based Statistics*. 2nd ed., revised by Tim Erickson. Indianapolis: John Wiley & Sons, 2004. (Specifically, see the activities "What Is Random Behavior," "What's the Chance?" and "Dueling Dice.")

Shaughnessy, J. Michael, Gloria Barrett, Rick Billstein, Henry A. Kranendonk, and Roxy Peck. *Navigating through Probability in Grades 9–12. Principles and Standards for School Mathematics* Navigations Series. Reston, Va.: National Council of Teachers of Mathematics, 2004.

Algebra 2 resources

Burrill, Gail, Christine A. Franklin, Landy Godbold, and Linda J. Young. *Navigating through Data Analysis in Grades 9–12. Principles and Standards for School Mathematics* Navigations Series. Reston, Va.: National Council of Teachers of Mathematics, 2003.

Koban, Lori, and Erin McNeils. "Fantasy Baseball with a Statistical Twist." *Mathematics Teacher* 102 (November 2008): 264–71.

Peck, Roxy, and Daren Starnes. *Making Sense of Statistical Studies*. Alexandria, Va.: American Statistical Association, 2009.

Scheaffer, Richard L., Ann Watkins, Jeffrey Witmer, and Mrudulla Gnanadesikan. *Activity-Based Statistics*. 2nd ed., revised by Tim Erickson. Indianapolis: John Wiley & Sons, 2004. (Specifically, see the activities "Random Rectangles," "Streaky Behavior," "Spinning Pennies," and "How Accurate Are the Polls?")

Tarr, James E., Hollylynne Stohl Lee, and Robin L. Rider. "When Data and Chance Collide: Drawing Inferences from Empirical Data." In *Thinking and Reasoning with Data and Chance,* Sixty-eighth Yearbook of the National Council of Teachers of Mathematics (NCTM), edited by Gail F. Burrill, pp. 139–49. Reston, Va.: NCTM, 2006.

Teague, Daniel J. "Experimental Design: Learning to Manage Variability." In *Thinking and Reasoning with Data and Chance,* Sixty-eighth Yearbook of the National Council of Teachers of Mathematics (NCTM), edited by Gail F. Burrill, pp. 151–69. Reston, Va.: NCTM, 2006.

Appendix 3
Tasks

This book discusses the use of rich tasks in the classroom to bring to the surface students' understandings and misunderstandings about statistics. Three sample tasks are presented here for the reader's personal or classroom use. At nctm.org /more4u, these tasks appear in an $8^{1}/_{2}$-by-11 format with templates for hands-on use with students.

Inventing Activities

Notes for the teacher: As indicated in Chapter 2, *inventing activities* are designed to "help students learn from the direct instruction that is likely to occur at other points in the class" (Schwartz and Martin 2004, p. 130). In fact, Schwartz and Martin (2004) propose that inventing activities are somewhere between direct instruction and discovery models of teaching and serve as "a counterexample to the misconception that direct instruction is against constructivist principles and should therefore be avoided" (p. 131). They claim that an inventing activity "alleviates the burden of carefully guiding students to discover the correct solutions—the teacher can simply explain the solution after the students have been 'prepared'" (p. 131).

Two activities developed by Roll (2009) illustrate common features of inventing activities. Both involve quality-assurance tests of products. The first is set in a commercial context: a trampoline manufacturer is running rebound tests on its products to ensure that they have the requisite bounce. The second is set in an aerospace context: NASA is testing rockets to ensure that they meet the agency's primary specification—that they consistently and reliably travel a distance of 500 miles.

In the trampoline activity, students compare two data sets that differ with respect to a key feature in this case—variability. The two sets have equal means and sample sizes but different ranges. After your students attempt to invent a method for comparing the trampolines, provide them with direct instruction on known procedures—for example, the process for computing the mean absolute deviation.

Trampoline Testing

The Bouncers Trampoline Company tests its trampolines by dropping a 100-pound weight from a height of 15 feet. The testers measure how many feet the weight rebounds into the air. They perform several trials on each trampoline. Figure 1 presents the results for two of the company's trampolines.

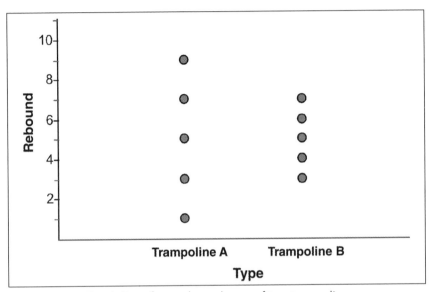

Fig. 1. Data from rebound tests of two trampolines

Consider the data on rebound heights for each of the two trampolines, A and B. For which trampoline are the heights closer to some middle (or average) value? Create a method for determining which trampoline's rebound heights are closer to a single point. You should use the same method to evaluate both trampolines. Your method should result in a single value for each trampoline. Write your method in steps so that others can apply it.

From "Structured Invention Tasks to Prepare Students for Future Learning: Means, Mechanisms, and Cognitive Processes" by Ido Roll (PhD dissertation, Carnegie Mellon University, Pittsburgh, Pa., 2009).

Trampoline Testing

Follow-up for the teacher: The data sets in figure 1 differ only in their standard deviations. A useful exercise might be to have students explore data sets that differ in other key ways as well. The data sets in figure 2 have the same sample sizes, means, and ranges, but different standard deviations. The data sets in figure 3 have the same means and ranges, but different sample sizes and standard deviations. Remember that students may invent novel methods for comparing the variabilities of the two data sets, but the goal for this activity is to pave the way for the direct instruction that is to come.

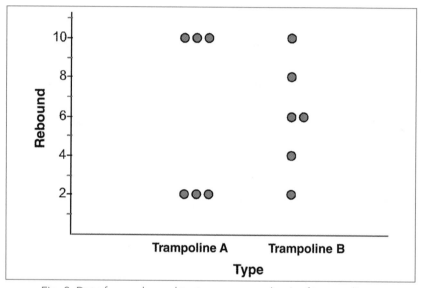

Fig. 2. Data from rebound tests on a second pair of trampolines

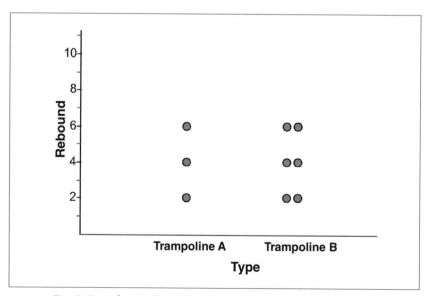

Fig. 3. Data from rebound tests on a third pair of trampolines

Rocket Testing

NASA needs to choose a rocket to launch its latest satellite. No rocket is currently ready, but NASA wants to choose one design and focus on its development. At this point, NASA does not care about the absolute height that the rocket reaches, since the amount of fuel can be adjusted. NASA cares about the ability to predict that the rocket will reach a target height of 500 miles. The agency needs a rocket that will reach, or come close to, the target height every time. Figure 1 shows the height that rockets of four types reached during testing, relative to the desired height. Each point represents the height that a rocket reached in a single test. Which type do you think is the most appropriate rocket for this task? Evaluate all four rocket types.

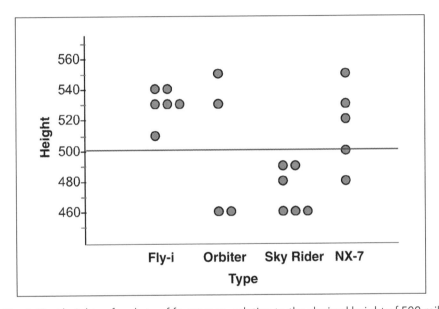

Fig. 1. Test heights of rockets of four types, relative to the desired height of 500 miles

Create a method for assessing the data from the rockets of four types and determine which rocket is likely to suit NASA's goals on the basis of the data for that rocket. You should use the same method to evaluate all four types of rockets. Your method should give a single value for each rocket type. Write your method in steps so that other people can apply it.

From "Structured Invention Tasks to Prepare Students for Future Learning: Means, Mechanisms, and Cognitive Processes" by Ido Roll (PhD dissertation, Carnegie Mellon University, Pittsburgh, Pa., 2009).

Developing an Estimator

Notes for the teacher: As discussed in Chapter 4, using a statistic to estimate an unknown parameter of a population under study involves understanding the roles of bias, variability, and random sampling in estimators. Developing an estimator can be challenging for students as a result. Chapter 4 introduces the activity Raffling to Support Recycling, which is set in the context of a raffle held by the Ecology Club at Red Mountain High School. The activity is designed for students working in groups of three or four with raffle tickets that can be made from a template at nctm.org/more4u. Each group chooses seven tickets at random from a population of N sequentially numbered tickets, and the students consider possible ways to estimate N on the basis of their sample of size 7.

As detailed below, the setup for the activity involves providing bags of N raffle tickets to students working in small groups. A template for raffle tickets simplifies the process.

1. Divide the class up into groups of three or four students.

2. Using the Raffle Tickets template at More4U, make a set of N sequentially numbered tickets for each group. Printing the tickets on cardstock gives sturdy tickets.

   Tickets on the template are numbered from 01251 to 01422, making $N = 172$ the maximum population of consecutively numbered tickets possible with the template. As an alternative to printing tickets, you can obtain rolls of sequentially numbered tickets from many office supply stores and online retailers. However, tickets must be numbered 01251 to 01422 for you to make classroom use of the sampling distribution results presented in the follow-up notes for the teacher (see the section **Sample simulation results**).

   Tickets numbered from 02151 to 01422 are also necessary if your students are to consider the seven numbers held by Jayla, her five friends, and Kevin as an additional random sample. Otherwise, Jayla's data cannot be used meaningfully in step 6 of the activity.

3. Give each group a brown paper bag containing N sequentially numbered tickets. Each bag should have the same number of tickets, ideally with the same starting and ending numbers. However, the only information that the students should have is that the tickets are sequentially numbered. They should *not* know the lowest ticket number.

4. Introduce the Raffling to Support Recycling scenario, and distribute the activity sheet.

Raffling to Support Recycling

Scenario

The Ecology Club at Red Mountain High School is holding a raffle to raise money for a field trip to the local recycling center. The prize is a $100 gift certificate to a local movie multiplex and was donated by the theater. The club is using a roll of raffle tickets left over from last year's fundraiser, and the tickets are numbered sequentially.

Jayla and each of five friends have purchased a $1 ticket at different times over the last two weeks from the club's booth near the principal's office. They are excited about the possibility of winning the gift certificate and would like to know what their chances are. Meghan points out that if 120 tickets were sold, each of them would have a chance of 1 out of 120. So if they knew that N tickets were sold, they could calculate the chances of one ticket being the winner as $1/N$, and the chances of one of the six of them being a winner as $6/N$. Their ticket numbers are

Kevin buys his ticket later than Jayla and her friends, and his number is 01419, but tickets are still available for sale after Kevin's purchase. When Kevin tells Jayla and her friends what his ticket number is, they realize that tickets bought later have higher numbers.

Suppose that Jayla and her friends do not know the number of the first ticket sold. They know only their own numbers and Kevin's number, and that the ticket numbers are sequential. How can they use these data to estimate N?

Activity

You have been assigned to a group, and your group has been given a bag of consecutively numbered tickets:

1. Randomly draw seven tickets out of the bag without replacement. Do not look in the bag while doing so. Record your random sample of seven ticket numbers below.

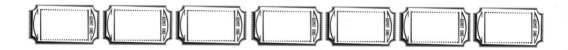

2. Discuss in your group how you could use these data (only these seven ticket numbers) to estimate N, the total number of tickets in the bag. Remember that the ticket numbers do not necessarily start with 00001.

3. How can you know whether your estimate of N is a good one?

4. Describe a method (or give a formula) for estimating the total number of tickets, N, based on a sample of size 7. Your method should be general enough so that it could be applied to any sample of seven tickets drawn without replacement from the bag.

5. Record your sample of seven ticket numbers on the board along with the other groups' samples of seven ticket numbers.

6. Apply your method of estimating N to the samples drawn by the other groups and to Jayla's data.

7. Your teacher will tell you how many tickets are in the bag. Which group's method gave the best estimate?

8. If your method came closest to estimating the true value of N here, does that mean that your method will always outperform other methods at estimating N?

9. How would you evaluate which method is the best one to use in general?

Raffling to Support Recycling

Follow-up for the teacher: The following information is provided to support discussion with students after they have completed the activity.

Talking points

Discussion after your students have completed the activity might address the following questions:

1. On the basis of the students' results, discuss as a class which method is the "best" for estimating *N*.

2. What does "best" mean?

3. Which method would give the most accurate estimate of *N*? How could this be tested?

4. How do *bias* and *variability* come into play in answers to the preceding questions?

   You may want to show your students some simulation results corresponding to methods of estimation that they have developed. Results from nine methods are shown below (see Sample simulation results). If you wish, you can incorporate these as part of a discussion of how simulation can be used to evaluate the properties of an estimator.

5. What properties of an estimation method would lead to an *unbiased* estimator?

6. Compare simulation results for the sampling distribution of various estimators. Which estimator seems preferable?

Methods of estimation

The table below shows nine different methods for estimating *N*, the number of tickets in the population, on the basis of the sample of Jayla and her friends' seven tickets. You may decide to consider these methods with your students; the methods reflect the nine histograms shown below (see "Sample simulation results") and can be paired with them, one to one, in order.

Method	Formula	Estimate for Jayla's data
1. Sample range	$\hat{N}_1 = \text{max} - \text{min}$	156
2. Four times the sample standard deviation	$\hat{N}_2 = 4s$	256
3. Six times the sample standard deviation	$\hat{N}_3 = 6s$	384
4. Four times the sample interquartile range	$\hat{N}_4 = 4IQR$	504*
5. Modified sample range	$\hat{N}_5 = \dfrac{k+1}{k-1}(\text{max} - \text{min}) - 1$	207
6. Modified sample standard deviation	$\hat{N}_6 = s\sqrt{12}$	222
7. Sample maximum	$\hat{N}_7 = \text{max}$	1419
8. Two times the sample mean	$\hat{N}_8 = 2\bar{x}$	2636
9. Two times the sample median	$\hat{N}_9 = 2m$	2578

* Fathom was used to compute the quartiles of the set of data. Different equally valid methods exist, and their use would lead to other results for these estimates.

The first four lines of the table show common methods of estimation that students might come up with for this problem. A perceptive student might realize that the range in a *sample* of the tickets sold cannot be larger than N, the actual number of tickets sold, and the sample range is likely to be an underestimate of N. A "better" estimate of N based on the sample range can be shown to be

$$\hat{N}_5 = \frac{k+1}{k-1}(\text{max} - \text{min}) - 1$$

where k is the number of tickets in the sample. For k = 7 in the activity, that estimate would be

$$\hat{N}_5 = \frac{8}{6}(\text{max} - \text{min}) - 1.$$

This estimator is listed in line 5 of the table. Line 6 gives an estimator that is a modification of the method based on taking a multiple of the sample standard deviation. Using the multiplier $\sqrt{12}$ is supported by a theoretical relationship between the range and the standard deviation of a uniform distribution.

An important feature of the Raffling to Support Recycling problem is that the lowest ticket number is not known. Students who do not realize this may propose methods for estimating N that have good properties in the situation where the ticket numbers begin with 00001 but may give unreasonably large estimates of N when the lowest ticket number is larger than 00001. The last three lines of the table give examples of this.

The "correct" answer

Although no estimate can be identified as the "correct" answer to the problem, N does have a true value that your students are trying to estimate. For the activity as conducted with raffle tickets made from the template supplied, the data are a random sample of 7 tickets drawn without replacement from a population of tickets numbered 01251 to 01422, and $N = 172$.

Possible modifications of the activity

The activity can be modified in various ways with implications that are useful for students to consider:

1. If the bags of tickets are set up so that each has the same number of tickets N, but in each bag the sequentially numbered tickets start and end with *different* values (for example, 01251 to 01422 in one bag, 01749 to 01920 in another bag, etc.), methods for estimating N that are based on the sample range, the standard deviation, or the interquartile range will be unaffected and still have the same properties shown in the graphs below. But methods that do not take into account the fact that the lowest ticket number is unknown will result in estimates of N that vary dramatically from group to group. In this situation, when your students compared their estimates, you could point out that all bags contained the same number of tickets.

2. If the tickets that you use for the activity are numbered sequentially beginning with 00001 and you make your students aware of that fact, then you will alter the activity to make it equivalent to the well-known German Tank problem. In that case, the "best" estimator is not one of the ones shown in the table above. An Internet search will turn up many references to this activity. For example, on the AP Statistics home page (http://apcentral.collegeboard.com/apc/public/courses/teachers_corner/2151.html), under "Special Focus Materials," click on Sampling Distributions.

Sample simulation results

The histograms below show paired simulation-based estimates of the sampling distribution for each of the methods in the table above, along with summary statistics related to the results. Each histogram is based on 50,000 simulations. In each graph, one of the vertical lines shows the location of the "true value" of N used in the simulations, $N = 172$, and the other vertical line is the location of the mean of the sampling distribution for that statistic. The relative accuracy and precision of each estimator is readily apparent in these graphs.

Although the Raffling to Support Recycling activity is similar to the German Tank estimation problem, it differs from that problem in that the raffle ticket numbers, unlike the tank serial numbers, do not necessarily begin with 00001. For this reason, estimation methods 7, 8, and 9 do not do well in the Raffling to Support Recycling problem.

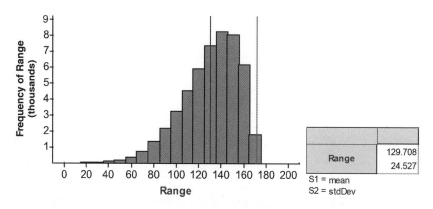

Range	129.708	
	24.527	

S1 = mean
S2 = stdDev

Method 1: Sample Range

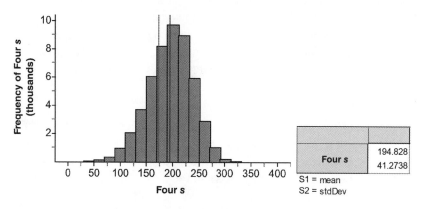

Four s	194.828	
	41.2738	

S1 = mean
S2 = stdDev

Method 2: Four times the sample standard deviation

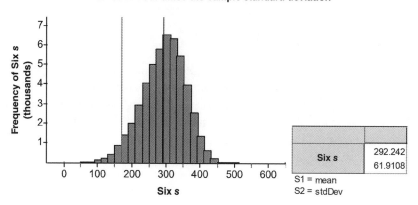

Six s	292.242	
	61.9108	

S1 = mean
S2 = stdDev

Method 3: Six times the sample standard deviation

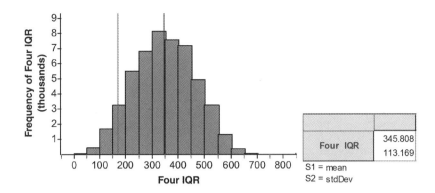

Four IQR	345.808	
	113.169	

S1 = mean
S2 = stdDev

Method 4: Four times the sample interquartile range.
(See note under table above regarding the computation of the
quartiles of a set of data.)

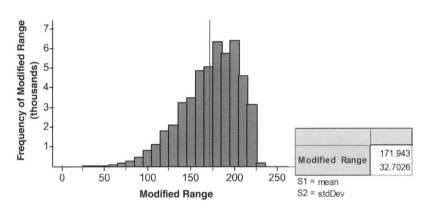

Modified Range	171.943	
	32.7026	

S1 = mean
S2 = stdDev

Method 5: Modified sample range

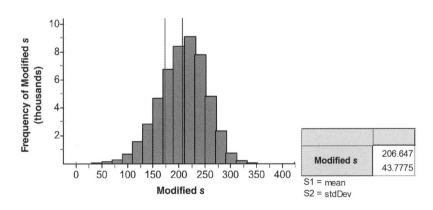

Modified s	206.647	
	43.7775	

S1 = mean
S2 = stdDev

Method 6: Modified sample standard deviation

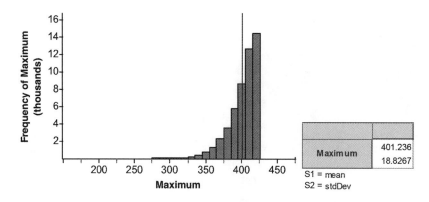

Method 7: Sample maximum

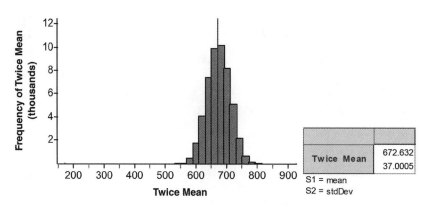

Method 8: Twice the sample mean

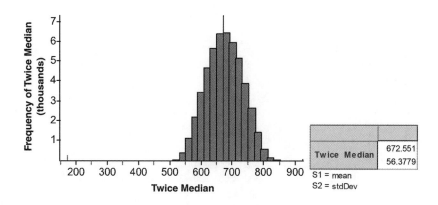

Method 9: Twice the sample median

References

Aliaga, Martha, George Cobb, Carolyn Curr, Joan Garfield, Rob Gould, Robin Lock, Tom Moore, et al. *Guidelines for Assessment and Instruction in Statistics Education: College Report.* Alexandria, Va.: American Statistical Association, 2005. http://www.amstat.org/education/gaise/GaiseCollege_Full.pdf.

Bower, Keith M. "Some Misconceptions about the Normal Distribution." Paper presented at the Six Sigma Forum, Milwaukee, Wis., 2003. http://www.minitab.com/uploaded-Files/Content/News/Published_Articles/normal_distribution_misconceptions.pdf.

Box, George E. P., and Norman R. Draper. *Empirical Model Building and Response Surfaces.* New York: John Wiley & Sons, 1987.

Carmona, Guadalupe, and Steven Greenstein. "Investigating the Relationship between the Problem and the Solver: Who Decides What Math Gets Used?" In *Modeling Students' Mathematical Modeling Competencies,* edited by Richard Lesh, Peter L. Galbraith, Christopher R. Haines, and Andrew Hurford, pp. 245–54. New York: Springer, 2010.

Chance, Beth, Robert delMas, and Joan Garfield. "Reasoning about Sampling Distributions." In *The Challenge of Developing Statistical Literacy, Reasoning and Thinking,* edited by Dani Ben-Zvi and Joan Garfield, pp. 295–323. Dordrecht, The Netherlands: Kluwer Academic, 2004.

Cobb, George W., and David S. Moore. "Mathematics, Statistics, and Teaching." *The American Mathematical Monthly* 104 (November 1997): 801–23.

College Board. AP Statistics: Course Description (2010). http://media.collegeboard.com/digitalServices/pdf/ap/ap-statistics-course-description.pdf.

delMas, Robert C., Joan Garfield, and Andrew Zieffler. "Using Model Eliciting Activities to Teach Statistics." Workshop presented at the Annual AMATYC Meeting, Las Vegas, Nev., November 14, 2009.

Dougherty, Barbara J. "Access to Algebra: A Process Approach." In *The Future of the Teaching and Learning of Algebra*, edited by Helen Chick, Kaye Stacey, Jill Vincent, and John Vincent, pp. 207–13. Victoria, Australia: University of Melbourne, 2001.

Fidler, Fiona, and Geoff Cumming. "Teaching Confidence Intervals: Problems and Potential Solutions." Paper presented at the International Statistical Institute, Fifty-fifth Session, Sydney, Australia, April, 2005. http://iase-web.org/documents/papers/isi55/Fidler-Cumming.pdf.

Garfield, Joan. "How Students Learn Statistics." *International Statistical Review* 63, no. 1 (1995): 25–34.

Garfield, Joan, and Dani Ben-Zvi. "How Students Learn Statistics Revisited: A Current Review of Research on Teaching and Learning Statistics." *International Statistical Review* 75, no. 3 (December 2007): 372–96.

Garfield, Joan, Robert delMas, and Andrew Zieffler. "Developing Tertiary-Level Students' Statistical Thinking through the Use of Model-Eliciting Activities." Paper presented at ICOTS8 (Eighth International Conference on Teaching Statistics), Ljubljana, Slovenia, July, 2010. http://iase-web.org/documents/papers/icots8/ICOTS8_8B3 _GARFIELD.pdf.

Grossman, Pamela. *The Making of a Teacher.* New York: Teachers College Press, 1990.

Hamilton, Eric, Richard Lesh, Frank Lester, and Michael Brilleslyper. "Model-Eliciting Activities (MEAs) as a Bridge between Engineering Education Research and Mathematics Education Research." *Advances in Engineering Education* 1, no. 2 (Summer 2008): 1–12. http://advances.asee.org/wp-content/uploads/vol01/issue02/papers /aee-vol01-issue02-p06.pdf.

Hancock, Chris, James J. Kaput, and Lynn Goldsmith. "Authentic Inquiry with Data: Critical Barriers to Classroom Implementation." *Educational Psychologist* 27, no. 3 (1992): 337–64.

Harradine, Anthony, Carmen Batanero, and Allan Rossman. "Students' and Teachers' Knowledge of Sampling and Inference." In *Teaching Statistics in School Mathematics—Challenges for Teaching and Teacher Education: A Joint ICMI/IASE Study: The 18th ICMI Study*, edited by Carmen Batanero, Gail Burrill, and Chris Reading, pp. 235–36. New York: Springer, 2011.

Hill, Heather C., Brian Rowan, and Deborah Loewenberg Ball. "Effects of Teachers' Mathematical Knowledge for Teaching on Student Achievement." *American Educational Research Journal* 42 (Summer 2005): 371–406.

Larson, Christine. "Modeling and Quantitative Reasoning: The Summer Jobs Problem." In *Modeling Students' Mathematical Modeling Competencies,* edited by Richard Lesh, Peter L. Galbraith, Christopher R. Haines, and Andrew Hurford, pp. 111–18. New York: Springer, 2010.

Lesh, Richard, Miriam Amit, and Roberta Y. Schorr. "Using 'Real-Life' Problems to Prompt Students to Construct Conceptual Models for Statistical Reasoning." In *The Assessment Challenge in Statistics Education*, edited by Iddo Gal and Joan B. Garfield, pp. 65–83. Amsterdam: IOS Press, 1997.

Lesh, Richard, Mark Hoover, Bonnie Hole, Anthony Kelly, and Thomas Post. "Principles for Developing Thought-Revealing Activities for Students and Teachers." In *Research Design in Mathematics and Science Education,* edited by Anthony Kelly and Richard Lesh, pp. 591–646. Mahwah, N.J.: Lawrence Erlbaum, 2000.

Magnusson, Shirley, Joseph Krajcik, and Hilda Borko. "Nature, Sources, and Development of Pedagogical Content Knowledge for Science Teaching." In *Examining Pedagogical Content Knowledge,* edited by Julie Gess-Newsome and Norman G. Lederman, pp. 95–132. Dordrecht, The Netherlands: Kluwer Academic, 1999.

Maloney, Alan P., Jere Confrey, Dicky Ng, and Jennifer Nickell. "Learning Trajectories for Interpreting the K–8 Common Core State Standards with a Middle–Grades Statistics Example." In *Using Research to Improve Instruction, Annual Perspectives in Mathematics Education* 2014, edited by Karen Karp, pp. 23–33. Reston, Va.: National Council of Teachers of Mathematics, 2014.

National Governors Association Center for Best Practices and Council of Chief State School Officers (NGA Center and CCSSO). *Common Core State Standards for Mathematics. Common Core State Standards (College- and Career-Readiness Standards and K–12 Standards in English Language Arts and Math).* Washington, D.C.: NGA Center and CCSSO, 2010. http://www.corestandards.org.

Peck, Roxy, Robert Gould, and Stephen J. Miller. *Developing Essential Understanding of Statistics for Teaching Mathematics in Grades 9–12.* Essential Understanding Series. Reston, Va.: National Council of Teachers of Mathematics, 2013.

Popham, W. James. "Defining and Enhancing Formative Assessment." Paper presented at the CCSSO State Collaborative on Assessment and Student Standards FAST meeting, Austin, Tex., October 10–13, 2006.

Reading, Chris, and J. Michael Shaughnessy. "Reasoning about Variation." In *The Challenge of Developing Statistical Literacy, Reasoning and Thinking*, edited by Dani Ben-Zvi and Joan Garfield, pp. 201–26. Dordrecht, The Netherlands: Kluwer Academic, 2004.

Roll, Ido. "Structured Invention Tasks to Prepare Students for Future Learning: Means, Mechanisms, and Cognitive Processes." PhD diss., Carnegie Mellon University, Pittsburgh, Pa., 2009. http://reports-archive.adm.cs.cmu.edu/anon/hcii/CMU-HCII-09-105.pdf.

Rossman, Allan J., and Beth L. Chance. "Teaching the Reasoning of Statistical Inference: A 'Top Ten' List." Washington, D.C.: Mathematical Association of America, 2000. http://rossmanchance.com/papers/topten.html.

———. "Anticipating and Addressing Student Misconceptions." Paper presented at ARTIST RoundTable Conference on Assessment in Statistics, Lawrence University, Appleton, Wis., August, 2004. http://www.rossmanchance.com/artist/proceedings/rossman.pdf.

Rossman, Allan, Beth Chance, and Elsa Medina. "Some Important Comparisons between Statistics and Mathematics, and Why Teachers Should Care." In *Thinking and Reasoning with Data and Chance*, Sixty-eighth Yearbook of the National Council of Teachers of Mathematics (NCTM), edited by Gail F. Burrill, pp. 323–33. Reston, Va.: NCTM, 2006.

Saldanha, Luis A. "'Is This Sample Unusual?' An Investigation of Students Exploring Connections between Sampling Distributions and Statistical Inference." PhD diss., Vanderbilt University, Nashville, Tenn., 2004.

Schwartz, Daniel L., and Taylor Martin. "Inventing to Prepare for Future Learning: The Hidden Efficiency of Encouraging Original Student Production in Statistics Instruction." *Cognition and Instruction* 22, no. 2 (2004): 129–84.

Seeley, Cathy L. "Constructive Struggling: The Value of Challenging Our Students." Message 17 in *Faster Isn't Smarter: Messages about Math, Teaching, and Learning in the 21st Century.* Sausalito, Calif.: Math Solutions, 2009. http://www.mathsolutions.com/documents/9781935099031_message17.pdf.

Shaughnessy, J. Michael. "Research on Students' Understanding of Some Big Concepts in Statistics." In *Thinking and Reasoning with Data and Chance*, Sixty-eighth Yearbook of the National Council of Teachers of Mathematics (NCTM), edited by Gail F. Burrill, pp. 77–98. Reston, Va.: National Council of Teachers of Mathematics, 2006.

———. "Research on Statistics Learning and Reasoning." In *Second Handbook of Research on Mathematics Teaching and Learning,* edited by Frank K. Lester, Jr., pp. 957–1009. Charlotte, N.C.: Information Age; Reston, Va.: National Council of Teachers of Mathematics, 2007.

Shulman, Lee S. "Those Who Understand: Knowledge Growth in Teaching." *Educational Researcher* 15, no. 2 (1986): 4–14.

———. "Knowledge and Teaching." *Harvard Educational Review* 57, no. 1 (1987): 1–22.

Sotos, Ana Elisa Castro, Stijn Vanhoof, Wim Van den Noortgate, and Patrick Onghena. "Students' Misconceptions of Statistical Inference: A Review of the Empirical Evidence from Research on Statistics Education." *Educational Research Review* 2, no. 2 (2007): 98–113.

———. "How Confident Are Students in Their Misconceptions about Hypothesis Tests?" *Journal of Statistics Education* 17, no. 2 (2009). www.amstat.org/publications/jse/v17n2/castrosotos.html.

Watson, Jane M., and Jonathan B. Moritz. "Developing Concepts of Sampling." *Journal for Research in Mathematics Education* 31 (January 2000): 44–70.

Wild, Chris J., and Maxine Pfannkuch. "Statistical Thinking in Empirical Enquiry." *International Statistics Review* 67 (December 1999): 223–48.

Wiliam, Dylan. "Keeping Learning on Track: Classroom Assessment and the Regulation of Learning." In *Second Handbook of Research on Mathematics Teaching and Learning,* edited by Frank K. Lester, Jr., pp. 1053–98. Charlotte, N.C.: Information Age; Reston, Va.: National Council of Teachers of Mathematics, 2007.

Yinger, Robert J. "The Conversation of Teaching: Patterns of Explanation in Mathematics Lessons." Paper presented at the meeting of the International Study Association on Teacher Thinking, Nottingham, England, May, 1998.